Intelligent Safety and Reliability Analysis of Rail Transit Trains

Yong Qin · Zhi-Peng Wang · Lin-Lin Kou ·
Xue-Jun Zhao · Yong Fu · Li-Min Jia

Intelligent Safety and Reliability Analysis of Rail Transit Trains

国防工業出版社
National Defense Industry Press

Springer

Yong Qin
Beijing Jiaotong University
Beijing, China

Lin-Lin Kou
Beijing Mass Transit Railway Operation Corporation Ltd.
Beijing, China

Yong Fu
Zhuzhou CRRC Times Electric Co.
Changsha, China

Zhi-Peng Wang
Beijing Jiaotong University
Beijing, China

Xue-Jun Zhao
CRRC Academy
Beijing, China

Li-Min Jia
Beijing Jiaotong University
Beijing, China

ISBN 978-981-97-4169-4 ISBN 978-981-97-4170-0 (eBook)
https://doi.org/10.1007/978-981-97-4170-0

Jointly published with National Defense Industry Press
The print edition is not for sale in China (Mainland). Customers from China (Mainland) please order the print book from: National Defense Industry Press.
ISBN of the Co-Publisher's edition: 978-7-118-12432-3

Translation from the English language edition: "Gui Dao Jiao Tong Lie Che An Quan Ke Kao Xing Fen Xi" by Yong Qin et al., © National Defense Industry Press 2022. Published by National Defense Industry Press. All Rights Reserved.

This Springer imprint is published by the registered company Springer Nature Singapore Pte Ltd.
The registered company address is: 152 Beach Road, #21-01/04 Gateway East, Singapore 189721, Singapore

Contents

Chapter 1
Introduction

Through the establishment and execution of national initiatives including the "Strong Transportation Nation" and the "National Comprehensive Three-Dimensional Transportation Network Planning Outline," rail transportation, serving as the fundamental infrastructure, has emerged as a robust assurance for facilitating the seamless realization of national strategies. More specifically, the high-speed railway system is indispensable in supporting China's economic and social development and enhancing its international social status. By the end of 2021, China's high-speed railway had attained 40,000 km of operation, forming the world's largest high-speed railway infrastructure network, with a national fleet of 4,153 standard sets and 33,221 vehicles, annually transporting 2,612 billion passengers; the operational urban rail transit network in China spans a total length of 7,253.7 km, servicing collective passenger traffic of 23.602 billion across 43 cities equipped with subway systems. Foreign-developed countries likewise place substantial emphasis on the advancement of the rail transit system and position it as a green and safe international competitive land mass transit infrastructure with "zero carbon and zero death."

The requisite for the regular functioning of rail transport systems and a fundamental element of their core competitive advantage is ensuring safety. The national strategy "Outline for the Construction of a Strong Transportation Country" issued by the Central Committee and the State Council has prioritized safety during the development of contemporary transportation systems and has proposed additional criteria for flawless and dependable safety assurance, rapid response, and enhancement of essential safety levels. Currently, the overall operational safety level of China's rail transit system is in a first-class international position, and yet with the expansion of the scale of operation of China's rail transit system and the increase in the service life of equipment, its operational safety is confronting significant challenges including deterioration of the service health of equipment and facilities, frequent disasters in the operating environment along the line, the errors in management and operations by site operators, as well as additional risks and hazards, including the lack of effective countermeasures will contribute to interruptions in train operation and stopping,

Y. Qin et al., *Intelligent Safety and Reliability Analysis of Rail Transit Trains*,
https://doi.org/10.1007/978-981-97-4170-0_1

foreign body intrusion affecting traffic and the lack of effective countermeasures will result in severe accidents including train interruptions and stops, foreign body intrusion affecting traffic and collisions with construction workers. Moreover, high-density operations, complex technical systems, the substantial interdependencies among various components, and the challenging environmental conditions present a considerable challenge to the safety of rail transit train operations. The comprehensive system to strengthen the global preventive safety of rail transit train operations and promote the safety assurance mode from traditional experience following the fact to active safety and autonomous safety assurance has emerged as a significant trend in the expansion of China's rail transit system. Currently, the primary objective in establishing a robust transportation infrastructure for a nation is centered on safety. Within the realm of science and technology innovation in rail transportation, specific directives include "advancing the utilization of intelligent detection and monitoring systems for traffic equipment, as well as operation and maintenance technology" and "elevating the standard of fundamental safety." As a successful example of China's independent innovation, China's high-speed railway is currently leading the world and will be developed even more during the 14th Five-Year Plan. Consequently, the theoretical and technical system of rail transit train system safety has emerged as a significant area of investigation and a hot issue for the future development of rail transit construction and operation in China.

As rail transport technology advances globally, the technical configuration of rail transport trains is progressively intricate, encompassing over 40,000 diverse subsystems and components with robust interdependencies. It represents a characteristic complex mechanical and electronic information system. The daily operation of these trains, occurring under demanding conditions such as high speed, high density, heavy load, and intense impact, results in rapid and stochastic nonlinear changes in equipment service performance. It also frequently faces the influence of external environmental factors including natural disasters and human errors; thus bringing great uncertainty and even catastrophic consequences to the safe operation of trains. For instance, on June 3, 1998, while traveling from Munich to Hamburg in Germany, the ICE 884 experienced the most catastrophic incident in the annals of high-speed rail. A malfunctioning wheelset component set off a chain of equipment failures that ultimately resulted in the train's high-speed derailment and the tragic loss of 101 lives among members of the passenger crew. On 6 November 2002, a high-speed train resulted in a fire in a sleeping car due to a short circuit in the train's electrical system, resulting in the death of 12 people. July 23, 2011, a railway traffic accident of notable severity transpired along the Yongwin Line, a high-speed railway in China., in which 40 people were killed and 172 injured due to the failure of the train's signaling equipment and the poor handling of subsequent risk emergencies. December 11, 2017, Sanyo, Tokaido, Japan the Shinkansen high-speed train "Hope 34" was determined to be a major potential accident on the Shinkansen urban rail line due to a crack in the bogie of the vehicle during high-speed travel, which resulted in gearbox oil leakage, scorched and blackened wheel joints, and was averted in time to avert severe repercussions. On March 23, 2017, Shenzhen Metro Line 1 was delayed due to a train door equipment failure, resulting in a delay of over 20 min in the east-bound direction from Baishi Zhou to the airport. A malfunctioning piece of metro train equipment on Line 1 caused trains to be delayed by over 20 min, impeding the orderly operation of the city and stranding a significant number of individuals at the metro station. As demonstrated by the aforementioned examples, the deterioration of the health status of train equipment has a huge impact on the safe operation of rail transportation, and its safety and reliability are simultaneously impacted by a multitude of intricate uncertainties including internal factors including equipment composition and operating conditions, as well as external factors namely, natural disasters and human errors.

The rail transit system constitutes a sophisticated engineering framework wherein human, machine, environment, and management intricately interact. Its safety and reliability are integral throughout the entire life cycle, encompassing planning, design, construction, operation, and maintenance. Meanwhile, with the development of new enabling technologies including modern information and communication and artificial intelligence, the intelligent features of rail transit systems are becoming ever more apparent, and the huge amount of real-time safety monitoring data offers additional resources and methods to enhance safety assurance capabilities.

The future development vision of "Zero causality" for the operational safety of railway systems has been put forward by foreign developed railway countries, and it is expected that new technologies including advanced sensors, the Internet of Things, artificial intelligence, and big data will be employed to effectively respond to the inherent risks of equipment and facilities as well as the stochastic dangers introduced by the external environment and operators. Besides, the European Union has proposed the implementation of the largest rail transport project in the world. The European Union has initiated Shift2Rail, the most extensive innovation initiative in the railway industry, comprising Safe4Rail for real-time monitoring of train safety, GoSAFERAIL for dynamic assessment and system risk awareness, and MOMIT for multidimensional surveillance of infrastructure and natural calamities. Initial research has been conducted in the areas of all-around condition awareness acquisition, equipment service health analysis, dynamic risk assessment, and highly intelligent/autonomous O&M and emergency response decision support, with an emphasis on mainline and high-speed rail for the purposes of validation and demonstration. To seize the leading international position in this field and enhance international competitiveness.

For the purpose of maintaining China's international top level in this technology and to confront the difficulties presented by the dangers and risks of complex rail transit systems, there is an urgent need to research a new generation of rail transit safety and security systems with higher safety objectives, wider sensing range, more comprehensive risk identification, enhanced intelligent processing, as well as a higher level of resilience - a highly autonomous level safety and security system. By showcasing various modes of rail transport through technological demonstrations, we aim to stimulate technological advancements in rail transport security and safety, thereby contributing to the development and operation of a robust transportation nation. Consequently, modern rail transit train safety and reliability research are more

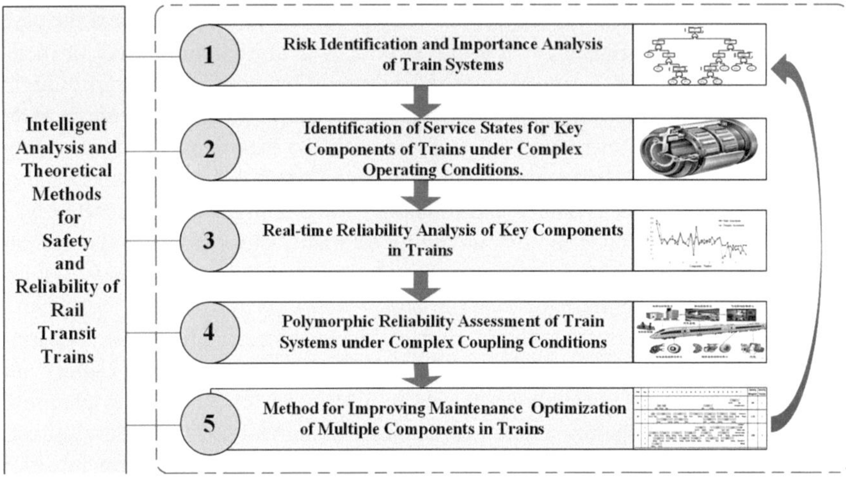

Fig. 1.1 Closed-loop processing framework for intelligent analysis of rail train safety and reliability

concerning strengthening the real-time, personalized, polymorphic, complex correlation and health state enhancement and optimization processing capabilities of safety and reliability analysis regarding train system risk assessment, train key component service state identification, train key component real-time reliability analysis, train complex system reliability calculation and train multi-part state repair as well as maintenance optimization, etc. For rail train dependability and safety, the system is founded upon a novel intelligent analysis method driven by the fusion of data and models. The system is a closed-loop unified processing framework consisting of five key components: risk assessment—service state identification—real-time reliability analysis—system multistate reliability calculation—multi-part state repair, as demonstrated in the following Fig. 1.1.

Specifically, these include

(1) Train system risk assessment and significance analysis. The entirety of the train system represents a sophisticated electromechanical information system, comprised of numerous components. It exhibits mutual coupling, time-varying, and random nonlinear characteristics, further influenced by the operational environment and numerous factors. It necessitates the exploration of analytical methods capable of comprehensively addressing the interdependencies, dual uncertainty, and dynamic attributes of comprehensive risk factors associated with the train system. Additionally, there is a requirement to employ train usage data for a more in-depth analysis of operational risks and their consequences.
(2) Service state identification of key train components in conditions of intricate operation. The failure process of key components of trains is subject to the combined effect of complex operating conditions and the external natural environment, and analyzing the extraction of signal characteristics from critical components in safe service while subjected to strong interference is an

urgent matter, non-linearity, and variable operating conditions and their state identification methods.

(3) Real-time reliability analysis and estimation of critical train components. Component performance degradation is positively correlated with the real-time dependability of critical train components. Accordingly, a high-precision key component performance degradation prediction model needs to be examined, and real-time reliability calculation is conducted on this basis.
(4) Multistate reliability analysis of complex train systems. On the basis of the quantitative analysis of the failure data of each component of the train system, the failure coupling mechanism between the components of the train system is illustrated, the reliability network model of the train system on the basis of the multistate complex network is constructed, and an algorithm for rapidly calculating the reliability of a complex system with multiple sources, sinks, and states is investigated.
(5) Train multi-part condition repair optimization and enhancement methods. To enhance the reliability of train operation, the comprehensive evaluation of the health status and maintenance relevance of a multi-part preventive maintenance strategy must be studied immediately, and an integrated strategy optimization model for multi-part preventive maintenance of trains as well as a fast solution algorithm for online maintenance optimization.

By delving into the theoretical approaches mentioned above, this research aims to elucidate the mechanisms governing the impact of safety risk factors on rail transit trains. It seeks to establish a theoretical and methodological framework for intelligent analysis of train safety and reliability, leveraging data and model fusion drive. The ultimate goal is to enhance the health status safety and reliability of train operations. This endeavor will provide robust theoretical underpinnings for the development of a new paradigm for analyzing and regulating train safety and reliability, actively contributing to ensuring proactive safety measures. In doing so, it is anticipated that this contribution will bolster and sustain China's preeminent position in this particular research domain.

Chapter 2
Operational Risk Analysis of Railway Train

2.1 Overview

In contemporary times, the global significance of the railway system has escalated, as it serves as a rapid, comfortable, and energy-efficient mode of transportation. This is particularly evident in China, where the railway passenger volume has surpassed 7 billion. Any latent risk associated with railway trains has the potential to lead to severe accidents, resulting in unacceptable financial and human losses. Hence, to better traveling safety of passengers, it is necessary to conduct goal-based risk control to monitor and detect the risks of trains more accurately. The train's risk can be subdivided into operational risk in the event of a delay and comprehensive risk of the system.

2.1.1 Comprehensive Risk of Railway Train

Many traditional approaches including fault tree analysis (FTA) [1], event tree analysis (ETA) [2], Bayesian analysis [3], failure mode and effect analysis (FMEA) [4, 5], and analytic hierarchy process (AHP) on the basis of expert judgments [6] have been utilized for train risk assessment. Nonetheless, the aforementioned research exclusively examines specific failure modes or qualitative analyses, neglecting the systematic, statistical, and dynamic characteristics of the railway train. As a complex electromechanical system with over 40,000 components, the operational risk of railway trains is involved with time-varying device states, component interconnections, changing working conditions, infrastructure, unpredictable environment, and staff aspects. Accordingly, an efficient modified risk assessment approach for railway trains is urgently in demand with consideration of systematicness, comprehensiveness, dual uncertainty, and dynamic nature. In addition, the primary objective of operational risk assessment for railway trains is to generate a rational and scientific

Y. Qin et al., *Intelligent Safety and Reliability Analysis of Rail Transit Trains*,
https://doi.org/10.1007/978-981-97-4170-0_2

ranking of component risks for the purposes of subsequent detection, risk management, and scheduling system maintenance. Thus, the railway train operational risk assessment on the basis of the maintenance schedule should consist of component risk sorting and multi-stage evaluation.

Among the distinct risk assessment technology, some papers find that the VlseKriterijuska Optimizacija I Komoromisno Resenje (VIKOR) approach can give more practical and applicable results. Meanwhile, in the domains of risk assessment and risk ranking, the VIKOR approach, which falls under the decision matrix technique and is a multiple-criteria decision-making method, can effectively manage calculation uncertainty, expert preference, and information indeterminacy. Fuzzy c-means clustering [7], trapezoidal fuzzy numbers [8], triangular fuzzy numbers [9, 10], Pythagorean fuzzy numbers [11], and intuitionistic fuzzy set (IFS) [12] have been incorporated into the VIKOR method to reckon with uncertainty in the risk assessment. By contrast, IFS can be highly useful in handling uncertainty and vagueness. Thus, the VIKOR approach with IFS remains applicable in railway train operational risk assessment in this study. Nonetheless, it can be challenging for decision-makers to get accurate values for the membership degree and the non-membership degree of IFS. Thus, a type-2 intuitionistic fuzzy set (type-2 IFS) [13] has been proposed to deal with these problems in a better way. Type-2 IFS possesses many advantages over type-1 IFS, as their membership function (MF) and non-membership function (NMF) are themselves fuzzy, enabling the modeling and reduction of indeterminacy's effects in fuzzy matters. Interval number IFS [4] is raised as an application of type-2 IFS to cope with fuzzy problems. Nonetheless, the value of upper MF and lower MF can be inaccurate sometimes, which can further enlarge the interval area after the interval arithmetic operation and affect the result. A triangular fuzzy number intuitionistic fuzzy set (TFNIFS) [14] has been proposed as another application of type-2 IFS. It can overcome the inaccurate interval area problem as the triangular fuzzy number between 0 and 1. Thus, the dual uncertainty problem might be better suited to TFNIFS. Despite some applications of TFNIFS in the fuzzy fields of science and technology [15], there are few applications for the railway train risk assessment yet. Therefore, it is praiseworthy to employ TFNIFS in this investigation to tackle the challenge of operational risk assessment for railway trains.

As discussed above, the mentioned studies just only discuss and analyze in a static circumstance. They ignore the dynamic risk properties in the assessment and cannot fully reflect the facts of the result to a certain extent. Railway train operational risk exhibits a dynamic risk characteristic, in addition to factors such as time-varying device states, variations in operational conditions, the prolonged and continuous work of staff, and the specialized knowledge of experts aimed at mitigating uncertainty-related risks. There are various applications of dynamic risk assessment in distinct fields. Meel and Seider [16] introduce dynamic failure assessment for the process facilities by updating the accident sequence probabilities. Moreover, Discrete-time Bayesian Network [17], dynamic Bayesian network [18], and dynamic fault tree analysis [19] are developed to assess the risk probabilities in the ever-changing environment. Furthermore, Qin et al. [20] present a dynamic risk assessment frame on the basis of the multiple stages to evaluate the status of metro stations. On the contrary, the aforementioned dynamic methods can be effectively utilized for short-term risk assessment predicated on the present condition. However, this is superfluous for the purpose of this paper and these methods cannot be combined with multiple risk criteria. The whole railway train operational risk assessment needs to be conducted on the basis of the train system and component maintenance.

2.1.2 *Operational Risk of Railway Train*

The advancement of national policy documents such as "Country's Rail and Road Networks" and "the Belt and Road Initiative" has ushered in a new phase in the development of China's high-speed railway infrastructure. By the end of 2019, China's high-speed railway operation mileage has reached 35000 km, and the basic high-speed railway network of "four vertical and four horizontal" has been preliminarily completed. Nonetheless, as the road network structure becomes more complex and transportation capacity resources become scarcer, equipment failure, bad weather, human interference, and additional risk factors have presented more formidable obstacles to the reliable functioning of high-speed trains. As one of the core competitiveness of high-speed railway brands, the guarantee of punctuality rate should be highly valued by China Railway Group.

As for the prediction research of train delay, the prediction models of train delay at this stage can be divided into static (on the basis of historical data) and dynamic (on the basis of real-time update data), deterministic, and stochastic-based (Taking interference between stations and sections into account) studies.

In recent years, numerous scholars have applied machine learning to prediction problems. Marković et al. [21] proposed a support vector regression prediction model on the basis of Serbian railway operational data, which can capture the relationship between the arrival delay of trains and various features of the railway system. Furthermore, the model utilizes seven input features pertaining to trains, schedules, infrastructure, and headways in order to produce predictions. It has been demonstrated that this model exhibits superior prediction accuracy in comparison to a conventional neural network model.

Oneto et al. [22] trained extreme learning machines as special cases of neural networks by machine learning methods on the basis of external dynamic data including train operational performance data and weather information of the Italian railway network. Its delay prediction system employed a fully data-driven approach to deal with large-scale networks, and adjusts it through a new threshold.

Francesco et al. [23] proposed a stochastic model on the basis of a Bayesian network to forecast the propagation of train delays on the busy trunk of Sweden using actual performance data of trains. This model obtained real-time information and updated the probability distribution to reduce the uncertainty of future train delay, which was an enhancement over the conventional prediction method on the basis of fixed values obtained offline from historical data. Moreover, in their study, Zeng et al. [24] developed a delay classification method and a delay propagation path construction form using operational performance data of high-speed trains from the dispatching section of the Beijing Railway Administration. To predict the time of train delays, they utilized a multi-layer perceptual neural network model. However, it is worth noting that the model's accuracy may be compromised in emergencies.

Moreover, Huang et al. [25] established a prediction model for the initial delay recovery of high-speed trains on the basis of the random forest algorithm under the assumption that the influencing factors of delay recovery remain unchanged by the operational performance data of Guangzhou high-speed trains. Nonetheless, this model did not take into account the increase in delay time arising from the consequences of delay that change with time and space. Moreover, Nair et al. [26] proposed a large-scale data-driven integrated prediction system for predicting train delays on the basis of the operational performance data of passenger trains in Germany. The system predicted train delay using a random forest model, a kernel regression model, and a simulation-based model. Among them, the simulation model takes into account the factors including the reserved buffer time of the train diagram, train connection, track occupation conflict, locomotive turnover plan, etc.

In practical transportation engineering problems, Bayesian networks are frequently employed in the fields of risk assessment, fault diagnosis, and prediction. Currently, the research by the Bayesian network predominantly focuses on fault diagnosis, while the research on the prediction of delay remains rare. Furthermore, Liang et al. [27] applied the Bayesian model on the basis of expert experience to the fault diagnosis of on-board train control equipment in the Wuhan Guangzhou Line. This model obtained fault symptom data from historical data and reduced attributes using rough set theory, thereby simplifying the training process. Zhao et al. [28] proposed a Bayesian network-based fault diagnosis method through information mining of fault records of on-board equipment of high-speed railway signal system, and obtained the optimal network structure through comparison of distinct structure learning algorithms (K2 algorithm/MCMC algorithm/expert knowledge). Ultimately, through a comparison with the outcomes of the KNN algorithm and BP algorithm, the superiority of this method was demonstrated.

Additionally, the current research on high-speed railway emergencies predominantly includes two aspects: the first is to analyze specific accidents by qualitative or semi-quantitative methods including expert experience, fault tree, tie map, etc. The second is to set up corresponding emergency response measures according to distinct classification standards. Nonetheless, there are few studies on the macro impact of emergencies on train operation on the basis of railway performance data, including the occurrence, distribution, and severity of emergencies. Understanding the correlation between emergencies and train operation can be achieved with greater objectivity through an examination of the space–time laws governing emergencies and their influence on train delays. It can also assist in developing a more reasonable and convincing scheduling adjustment plan, and avoid emergencies affecting the normal operation of trains.

2.2 Comprehensive Risk Analysis Method of Railway Train System Uncertainty

This chapter's objective is to establish a framework for the analysis of railway train system risks and the ranking of component risks, taking into account dual uncertainty, multiple risk criteria, and a multi-stage environment. In each stage, the risk criteria information can be updated by the changing risk factors and the deeper cognition from experts. Thus, a new risk ranking approach is proposed in this paper in conjunction with dynamic VIKOR and TFNIFS for the railway train system comprehensive risk analysis.

2.2.1 Railway Train Comprehensive Risk Analysis Index System

The purpose of the comprehensive risk analysis and key component identification of railway trains is to assess the risk of a multi-stage environment and rank the components. The operational risk of a railway train, which comprises more than 40,000 components, can be attributed to infrastructure, staff, time-varying device states, component interconnections, fluctuating working conditions, and an unpredictable environment. Thus, the comprehensive risk analysis and key component identification can be achieved by initially implementing an all-encompassing risk assessment index system and identifying key components with risk ranking by dynamic VIKOR and TFNIFS.

2.2.1.1 Index System

Various factors can affect the train operation. Thus, the risk factors can be divided into two main categories internal and external aspects.

In regard to internal risk factors, railway trains may be susceptible to the time-varying component states and their fluctuating working conditions. These elements may be denoted as the five risk criteria, including the mean distance between failure (MDBF), mean time to repair (MTTR), maintenance cost, fault detectability, and the risk effect on the system, people, and environment. Meanwhile, MDBF is another form of mean time between failure (MTBF), as time is replaced by running distance, and MDBF describes the probability of risk occurrence. Additionally, MTTR, maintenance cost, and the risk effect on the system, people, and environment of the train describe the severity of risk consequence, and MTTR measures the average time of the component repair. The expenditure associated with maintenance centers around the intricacy and importance of the system. The impact of risks on the system, individuals, and the environment is indicative of the severity of consequences resulting from the failure of system components.

Besides the inherent risk factors above, a railway train is a complex electromechanical system with over 40,000 interconnected components of distinct relationships and interactions, a minor malfunction in a single component has the potential to fail the entire train system. Hence, the component interconnections can be the correlation risk index of railway trains.

Infrastructure also has a substantial impact on the risk status of the railway train operation. Infrastructure factor involves various elements including tracks, bridges, electrical facilities, signal facilities, etc. In consideration of the factors mentioned earlier, it is imperative to acknowledge the direct correlation between the tracks and the operation of the train system. The quality of the tracks can affect the normal operation and safety status of railway trains as well as passenger comfort. Consequently, the track quality index (TQI) [29] can be employed to evaluate the risk of a railway train.

As a pivotal risk element, environmental factors exert a significant influence on the safety status of railway trains. Adverse weather conditions such as wind, heavy rain, snow, dust storms, and other inclement weather pose substantial challenges to the train system, potentially leading to performance degradation and functional failures. Thus, environmental factors based on extreme weather, temperature, and humidity should be taken into account for the train risk assessment.

The personnel factor stands out as one of the most critical and pivotal elements among the various factors affecting railway train operations. Train staff like train drivers and maintenance personnel are intimately tied to the operation of the train. It may cause large damage or even accidents if the train staff are inexperienced or fatigued. Consequently, the staff's operation skills and mental status can be applied to risk elements as staff factors. The operational skills of the staff can be comprehensively affected by age, education, and technical title. The mental status of the staff can be influenced by constant working periods.

As discussed above, the operational risk assessment index system of railway train for each component can be set up on the basis of the above factors, which can be indicated in Fig. 2.1. The above factors consist of quantitative and qualitative information, dividing the assessment criteria into numerical, interval, and fuzzy type. MDBF index, MTTR index, correlation risk index, operation skills index, mental status index, and TQI index can be calculated by certain specific formulas [30], and the result can be presented as numerical type with accurate numbers. Nonetheless, the risk effect index on system, people and environment, fault detectability index and environment risk index cannot be calculated by the accurate number and need to be represented by fuzzy set theory. Besides, the aforementioned factors can also be categorized into types based on their resultant impact, distinguishing between benefits and costs, and the detailed risk index information is demonstrated in Table 2.1.

2.2.1.2 Correlation Risk Index

A railway train is a complex electromechanical system consisting of more than 40,000 components, with various degradation and complex connection properties.

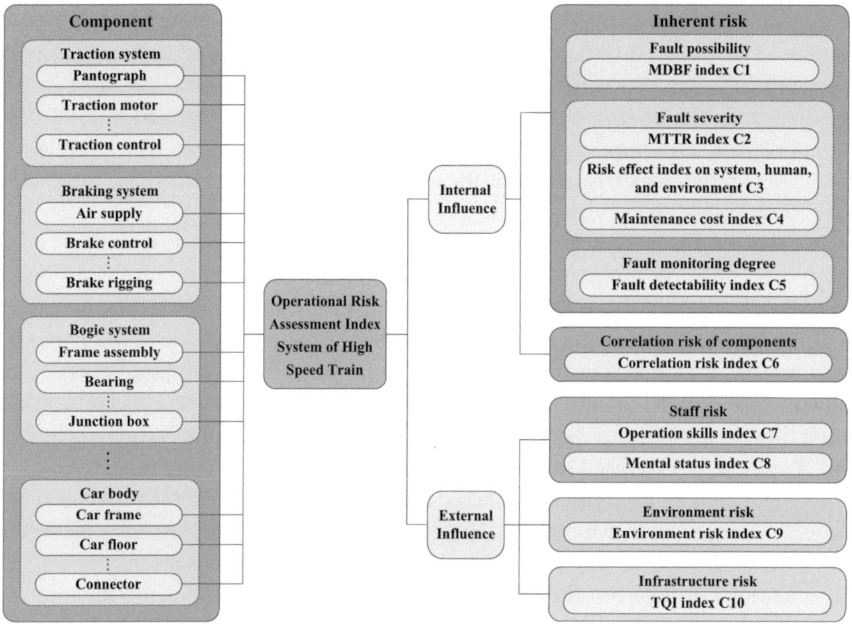

Fig. 2.1 Operational risk assessment index system of railway train

Table 2.1 Operational risk assessment index information

No.	Assessment index	Data type	Index type	Calculation formula
C_1	MDBF	Numerical	Benefit	$MDBF = \frac{\sum L}{N_f}$
C_2	MTTR	Numerical	Cost	$MTTR = \frac{\sum_{i=1}^{n} t_i}{n}$
C_3	Risk effect on system, people and environment	Fuzzy	Cost	–
C_4	Maintenance cost	Interval	Cost	$R_c = \frac{\bar{C}}{T}$
C_5	Fault detectability	Fuzzy	Benefit	–
C_7	Operation skill	Numerical	Benefit	$S_{rc} = \frac{\sum_{i=1}^{3} n_{1i}\varepsilon_i}{N} + \frac{\sum_{i=1}^{5} n_{2i}\lambda_i}{N} + \frac{\sum_{i=1}^{5} n_{3i}\eta_i}{N}$
C_8	Mental status	Numerical	Benefit	$Y_i = 1 + \sin\left(\frac{2\pi X}{23}\right)$
C_9	Environment risk	Fuzzy	Cost	–
C_{10}	TQI	Numerical	Cost	$TQI = \sum_{i=1}^{n} \sqrt{\frac{\sum_{j=1}^{n} (x_{ij} - \bar{x}_i)^2}{n}}$

Once a small failure occurs, it may lead to the system cascading failure and even train damage. There can be some defects in the internal risk factors analysis which only independently considers the component's properties and ignores the relationship between the components. Thus, the correlation risk index integrated with system structural risk and component failure is proposed in this paper. The correlation risk index in this paper is aimed at the interactions of components and their state importance in the train system.

In recent years, complex network theory based on the system structure has been utilized for the risk analysis of the relationships and interactions in the complex mechanical system. These researches focus on the structural properties including node degree and clustering coefficient on the basis of a single-layer network. Nonetheless, a single-layer network cannot represent the risk and complexity of railway trains with distinct mechanical, electrical, and information interactions. Accordingly, a multiplex network [31] coupled with numerous subnets of nodes and edges can be proposed to analyze the risk and complexity of railway trains.

Components in the railway train system contact and interact with each other through their respective functions. Nonetheless, a component may have divergent functions and result in distinct interactions. The loss of a function may not affect the relationships between the relative components, due to the distinct interactions between them. Consequently, the interactions of the components need to be divided according to the actual function of the railway train. The same function is placed on the same level, forming a mechanical, electrical, and information interaction layer of the train system. A multiplex network of railway trains is established and can be seen in Fig. 2.2.

In the multiplex network of railway trains, components and their interactions can be described as nodes and edges, respectively. Besides, the multiplex network of railway trains can be constructed by the mechanical interaction layer G1, electrical interaction layer G2, and information interaction layer G3 in accordance with

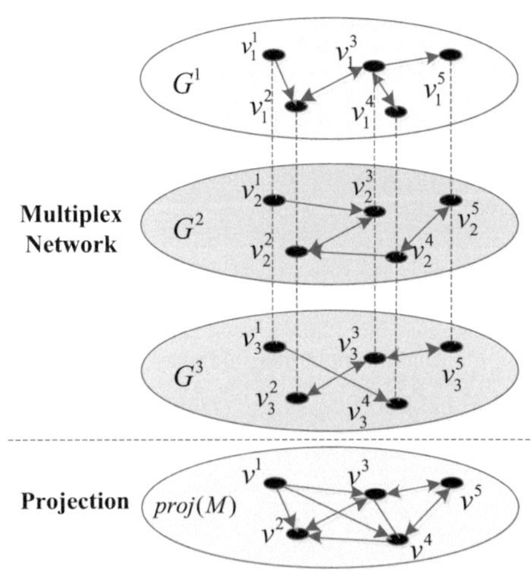

Fig. 2.2 Railway train multiplex network model

the actual function of the railway train. proj(M) is the projection of the multiplex network, which integrates all the interactions. Furthermore, in the mechanical interaction layer, components connect with each other by bolting and riveting, and the mechanical force can be transferred through the connections. Except for that, components in the electrical interaction layer are connected by cables and wires of distinct conductors that can transfer the electric energy. Moreover, components in the information interaction layer are connected through the transmission media which can transfer the command and state information. In the multiplex network, the identical component can integrate distinct information and functions through the layer connections. A directed edge can represent the relationship between distinct components. The one-way edge represents the one-way transmission, while the two-way edge represents the two-way transmission.

Multiplex network models can be always combined with the structural properties including node degree, clustering coefficient, closeness centrality, eigenvector centrality, etc. Nonetheless, every component of a train exhibits specific practical functional behaviors, influencing the overall risk status of the entire train system and demonstrating dynamic properties that evolve over time. Consequently, dynamic properties are proposed to analyze the correlation risk on the basis of the multiplex network structural properties and component failure status, as indicated in Table 2.2.

$K(i)$ serves as the failure degree of vi and is the result value on the basis of the topological degree and node failure synthetically. $K(i)$ represents the affected average range of neighbor nodes as the node fails. $C(i)$ denotes the failure clustering coefficient of vi, and measures the cluster degree after the node fails. $\bar{e}_{ij}$ indicates the number of the normal connections of vi after the node fails. $\bar{d}_{ij}$ symbolizes the length of the failure path for vi. $C_D(i)$ represents the failure closeness centrality of vi, and describes the centrality of vi as the node fails. $C_E(i)$ indicates the failure eigenvector centrality of vi and represents the importance of vi with the feedback of neighbor nodes. λ is the eigenvalue of its neighbor matrix. $I(i)$ is the value at risk of failure of vi and can reflect the network efficiency after the node fails. Moreover, E_0 and E_i are the network efficiency at the period of normal and failure operation, respectively.

The amalgamation of these five properties within the train multiplex network model, when coupled with component failures, imparts a dynamic characteristic to the railway train system. This framework proves more adept in depicting the

Table 2.2 Dynamic properties with failure of railway train multiplex network

Properties	Calculation formula
$K(i)$	$K(i) = \bar{\lambda}_i k_i = \sum_{j=1}^{m} \bar{\lambda}_i a_{ij}$
$C_D(i)$	$C_D(i) = (m-1) / \sum_{j=1}^{m} \bar{d}_{ij}$
$C_E(i)$	$C_E(i) = \frac{\sum_{j=1}^{m} \bar{\lambda}_i a_{ij} x_j}{\lambda}$
$I(i)$	$I(i) = 1 - \frac{E_i}{E_0}$

intricacies and risks associated with railway train operations. The correlation risk index can be an amalgamation of these five properties on the basis of the PCA approach [32]. Consequently, the property weight can be defined as

$$w_j = \frac{\sum_{j=1}^{n} a_j p_{ij}}{\sum_{j=1}^{n} \sum_{i=1}^{m} a_j p_{ij}} \tag{2.1}$$

where a_i indicates the eigenvalue of covariance matrix p with $n*m$ dimension, p_{ij} represents the principal component of the property C_j, and the correlation risk index of each component can be defined as

$$CR = \sum_{j=1}^{5} w_j C_j \tag{2.2}$$

To summarize, correlation risk index of railway train is predominantly aimed at the interactions and relationships of components as well as the state importance in the train system.

2.2.2 *Comprehensive Risk Analysis and Component Risk Ranking of Railway Train System Based on Type-2 Intuitionistic Fuzzy Set and Dynamic VIKOR Approach*

The operational risk assessment methodology of railway trains in this study has two major steps. The initial stage encompasses the establishment of a mixed fuzzy decision matrix for the assessment, while the subsequent step involves the application of dynamic VIKOR across multiple stages to derive the ultimate risk ranking of the train components. Figure 2.3 illustrates the procedure of the operational risk assessment methodology of railway trains.

2.2.2.1 Foundation of the Mixed Fuzzy Decision Matrix

Based on the railway train operational risk assessment index system, the risk criteria include both qualitative and quantitative information. As for the qualitative risk criteria like risk effect index on the system, people, and environment, fault detectability index, and environment risk index, they ought to be represented by fuzzy set theory rather than accurate numbers. TFNIFS [33] has been applied to cope with the dual uncertainty problem in this paper, as it possesses numerous advantages as its MF and NMF are themselves fuzzy. It is feasible to minimize the defect and error

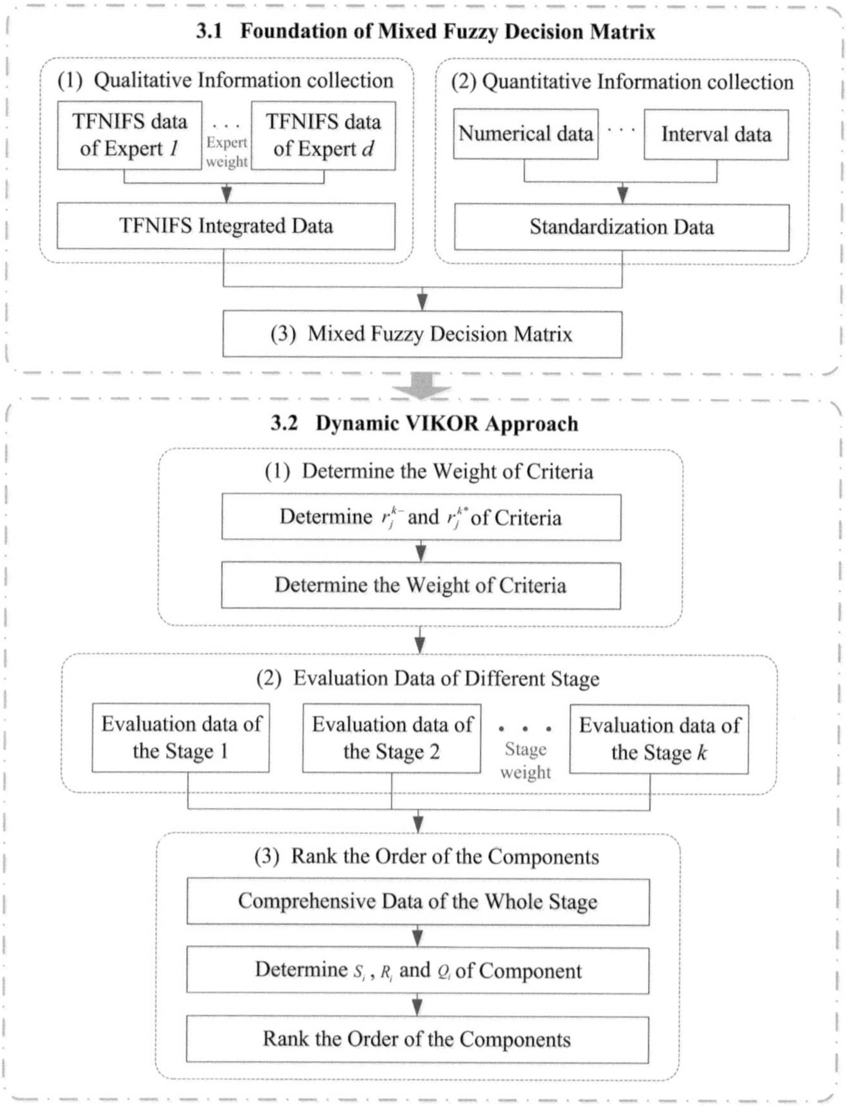

Fig. 2.3 Procedure of operational risk assessment of railway train on the basis of type-2 IFS and dynamic VIKOR approach

of the subjective judgments of experts and engineers. Consequently, TFNIFS can be more appropriate to describe the risk status of railway trains. Broadly speaking, the qualitative risk criteria within the TFNIFS can be articulated as

$$\vec{A} = \{\langle x, \mu_{\tilde{A}}(x), v_{\tilde{A}}(x)\rangle | \mathrm{x} \in \mathrm{X}\} \tag{2.3}$$

where $\mu_{\bar{A}}(x) = \left(\mu^1_{\bar{A}}(x), \mu^2_{\bar{A}}(x), \mu^3_{\bar{A}}(x)\right)$ and $\nu_{\bar{A}}(x) = \left(\nu^1_{\bar{A}}(x), \nu^2_{\bar{A}}(x), \nu^3_{\bar{A}}(x)\right)$ is MF and NMF, respectively, satisfying $\mu_{\vec{A}}(x)\colon X \to [0,1]$ and $\nu_{\vec{A}}(x)\colon X \to [0,1]$ expressed as triangular fuzzy numbers.

The study benefited from the active involvement of front-line experts and railway engineers in its execution, as they contributed by offering their subjective assessments of the qualitative risk criteria during the three assessment stages. ((al1, am1, ar1), (bl1, bm1, br1)) is a TFNIFS number $\vec{A}$, and is utilized as the value of expert evaluation. (al1, am1, ar1) and (bl1, bm1, br1) is the MF and NMF of $\vec{A}$, and represents the positive and negative degree of the risk criteria, respectively.

As for the risk effect index on the system, people, and environment, fault detectability index, and environment risk index of railway trains, the index linguistic variable described as TFNIFS type can be seen in Table 2.3. Take the environmental risk index as an example. Figure 2.4 shows TFNIFS with "Very Safe (VS)"—((0.0, 0.0, 0.1),(0.8, 0.8, 0.9)) in green, "Safe (S)"—((0.1, 0.2, 0.3), (0.6, 0.7, 0.8)) in blue, "Subcritical (0.4,0.5,0.6)) in purple, "Fatal (H)"—((0.5, 0.6, 0.7), (0.2, 0.3, 0.4)) in pink, and "Very Hazardous (VH)"—((0.7, 0.8, 0.9), (0.0, 0.1, 0.2)) in red, respectively. The solid lines describe the MF of TFNIFS, whereas the dotted lines describe the NMF of TFNIFS.

Later the weight of each expert can be determined by Formulas (2.6) and (2.7) on the basis of expert credibility [14].

$$B_s^k(\pi) = -\left[\left(\sum_{i=1}^{n}\sum_{j=1}^{m}\pi_{sij}^k\right)\ln\left(\sum_{i=1}^{n}\sum_{j=1}^{m}\pi_{sij}^k\right)\right]^{-1} \tag{2.4}$$

$$\lambda_s^k = \frac{B_s^k(\pi)}{\sum_{s=1}^{d} B_s^k(\pi)} \tag{2.5}$$

Table 2.3 Linguistic variable of qualitative criteria

TFNIFS	Linguistic variable C_3	Linguistic variable C_5	Linguistic variable C_9
((0.0,0.0,0.1), (0.8,0.9,0.9))	Slight (S)	Very Difficult (VD)	Very Safe (VS)
((0.1,0.2,0.3), (0.6,0.7,0.8))	Light (L)	Difficult (D)	Safe (S)
((0.3,0.4,0.5), (0.4,0.5,0.6))	Subcritical (SU)	Subcritical (SU)	Subcritical (SU)
((0.5,0.6,0.7), (0.2,0.3,0.4))	Fatal (F)	Easy (E)	Fatal (H)
((0.7,0.8,0.9), (0.0,0.1,0.2))	Disastrous (D)	Very Easy (VE)	Disastrous (VH)

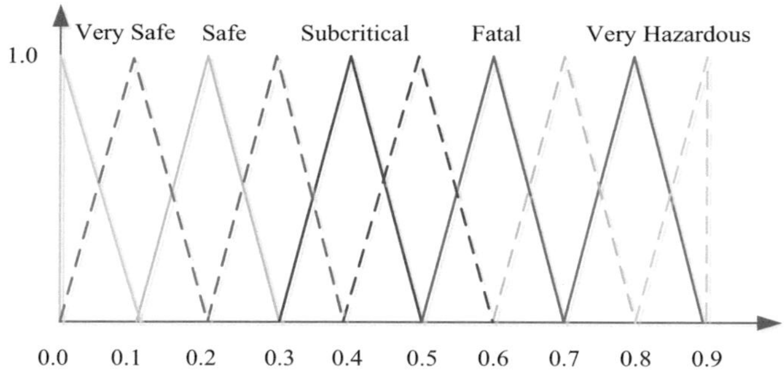

Fig. 2.4 Linguistic variable of environmental risk

where λ_s^k indicates the expert weight in the stage k, $B_s^k(\pi)$ symbolizes the expert credibility in the stage k, π_{sij}^k depicts the hesitancy degree of the qualitative evaluation criteria in the stage k. It exhibits an inverse correlation with the level of uncertainty inherent in the expert's assessment information.

$$\pi_{sij}^k = 1 - \mu_{sij}^k - v_{sij}^k \tag{2.6}$$

Apart from the qualitative criteria in the operational risk assessment, the quantitative risk criteria can be calculated based on Table 2.1. As for the criteria C4 (Maintenance Cost), it is an interval data, which can be expressed as $\left[a_1^L, a_1^U\right]$, a_1^L and a_1^U is the upper and lower bound of the data. Besides, the other criteria are the numerical data which can be expressed as an accurate number. Typically, the quantitative criteria data set UK needs to be standardized between 0 and 1 after the collection. Furthermore, Formulas (2.7) and (2.8) give an example of standardization for the criteria of cost and benefit type respectively.

$$r_{ij}^k = u_{ij}^k / \max u_{ij}^k \text{ or } r_{ij}^k = \left[u_{ij}^{kL} / \max u_{ij}^{kL},\ u_{ij}^{kU} / \max u_{ij}^{kU}\right] \tag{2.7}$$

$$r_{ij}^k = u_{ij}^k / \min u_{ij}^k \text{ or } r_{ij}^k = \left[u_{ij}^{kL} / \min u_{ij}^{kL},\ u_{ij}^{kU} / \min u_{ij}^{kU}\right] \tag{2.8}$$

where r_{ij}^k indicates the quantitative evaluation value after standardization, u_{ij}^k symbolizes the numerical evaluation value before standardization, u_{ij}^{kL} and u_{ij}^{kU} represent the upper and lower bound of the interval criteria before standardization.

Alternatives, evaluation criteria, and evaluation value including qualitative and quantitative information in the assessment stage can form the final mixed fuzzy decision matrix Rk. Qualitative evaluation information is given to the experts. The final mixed fuzzy decision matrix Rk can be represented as

$$R^k = \begin{array}{c} \\ A_1 \\ A_2 \\ \vdots \\ A_n \end{array} \begin{array}{c} \begin{array}{cccc} C_1 & C_1 & \cdots & C_1 \end{array} \\ \begin{bmatrix} r_{11}^k & r_{12}^k & \cdots & r_{1m}^k \\ r_{21}^k & r_{22}^k & \cdots & r_{2m}^k \\ \vdots & \vdots & \vdots & \vdots \\ r_{n1}^k & r_{n2}^k & \cdots & r_{nm}^k \end{bmatrix} \end{array} \tag{2.9}$$

$$W_j^k = \left[\omega_1^k, \omega_2^k, \cdots, \omega_m^k\right], \quad D_s = [D_1, D_2, \ldots, D_d] \tag{2.10}$$

where A_i ($A = \{A_1, A_2, \ldots, A_n\}$) is frequently applied to express the evaluation alternative (component), C_j ($C = \{C_1, C_2, \ldots, C_m\}$) is the evaluation criteria, D_s ($D = \{D_1, D_2, \ldots, D_d\}$) is the experts and r_{ij}^k is the comprehensive evaluation value including qualitative and quantitative information in the assessment stage K_k ($K = \{K_1, K_2, \ldots, K_v\}$), W_j^k is the weight of criteria C_j.

2.2.2.2 Dynamic VIKOR Approach

The VIKOR approach has been effectively applied in the domain of risk ranking and assessment based on the conclusive mixed fuzzy decision matrix Rk. It can determine the compromise ranking solution by a maximum group utility of the "majority" and a minimum individual regret of the "opponent", involving the preference of experts and engineers by index weights. Nonetheless, the VIKOR approach reckoned with the multiple criteria problems applied in a static circumstance, which cannot meet the requirements of dynamic operational risk assessment. Consequently, an extended. VIKOR approach on the basis of multiple stages is proposed for railway train operational risk assessment in this study.

Step 1: Determine the Weight of the Criteria in the Stage k

As for the multiple criteria in the final mixed fuzzy decision matrix Rk, each risk criterion is supposed to be compromised on the basis of the criteria weight and determined on the basis of entropy weight [34]. Prior to the criteria weight calculation, some intermediate parameters are supposed to be calculated. The initial components consist of the ideal and negative ideal solutions for the criteria, which must be identified within the decision matrix to facilitate subsequent calculations in the assessment process. They can be calculated on the basis of Formulas (2.11) and (2.12).

$$r_j^{k*} = \max_i r_{ij}^k, \; r_j^{k-} = \min_i r_{ij}^k \quad \text{for benefit type } C_j \tag{2.11}$$

$$r_j^{k*} = \min_i r_{ij}^k, \; r_j^{k-} = \max_i r_{ij}^k \quad \text{for cost type } C_j \tag{2.12}$$

The second intermediate parameter is the aggregation operator $\tilde{r}_j^k$ of criterion C_j, which can be demonstrated by $\tilde{r}_{Nj}^k$, $\tilde{r}_{Ij}^k$, and $\tilde{r}_{Tj}^k$ as numerical, interval, and TFNIFS data type according to Formulas (2.13), (2.14), and (2.15) respectively.

$$\tilde{r}_{Nj}^k = \left(r_{1j}^k + r_{2j}^k + \cdots + r_{3j}^k\right)/n \tag{2.13}$$

$$\tilde{r}_{Ij}^k = \left[\sum_{i=1}^{n} r_{ij}^{kL}/n, \sum_{i=1}^{n} r_{ij}^{kU}/n,\right] \tag{2.14}$$

$$\tilde{r}_{Tj}^k = \left(\left(\prod_{i=1}^{n}\left(a_{lij}^k\right)^{1/n}, \prod_{i=1}^{n}\left(a_{mij}^k\right)^{1/n}, \prod_{i=1}^{n}\left(a_{rij}^k\right)^{1/n}\right), \left(\left(1-\prod_{i=1}^{n}\left(1-b_{lij}^k\right)^{1/n}\right),\right.\right.$$
$$\left.\left.\left(1-\prod_{i=1}^{n}\left(1-b_{mij}^k\right)^{1/n}\right), \left(1-\prod_{i=1}^{n}\left(1-b_{rij}^k\right)^{1/n}\right)\right)\right) \tag{2.15}$$

The third intermediate parameter entropy e_j^k of criteria C_j can be calculated as

$$e_j^k = -\sum_{i=1}^{n}\left[d\left(r_{ij}^k, \tilde{r}_j^k\right)/\sum_{i=1}^{n} d\left(r_{ij}^k, \tilde{r}_j^k\right)\ln\left(d\left(r_{ij}^k, \tilde{r}_j^k\right)/\sum_{i=1}^{n} d\left(r_{ij}^k, \tilde{r}_j^k\right)\right)\right]/\ln(n) \tag{2.16}$$

where $d\left(r_{ij}^k, \tilde{r}_j^k\right)$ is the Euclidean distance between the r_{ij}^k and $\tilde{r}_j^k$.

Consequently, the weight of assessment criteria can be calculated as

$$\omega_j^k = 1 - e_j^k/\sum_{j=1}^{m}\left(1 - e_j^k\right) \tag{2.17}$$

Step 2: Evaluate the Risk Criteria Data from Distinct Stages

Concerning the entire duration of operational risk assessment, it is imperative to amalgamate the mixed fuzzy decision matrix Rk from each stage as comprehensive risk decision data, determined by the stage weight. The stage weight can also be computed using the entropy weight [35] in accordance with Formulas (2.18)–(2.20).

$$\bar{r}^k = E\left[\sum_{j=1}^{m}\left(r_{ij}^k/m\right)\right] \tag{2.18}$$

$$e^k = -\frac{\sum_{j=1}^{m}\left[\frac{d\left(\bar{r}_j^k,\bar{r}^k\right)}{\sum_{j=1}^{m} d\left(\bar{r}_j^k,\bar{r}^k\right)}\ln\left(\frac{d\left(\bar{r}_j^k,\bar{r}^k\right)}{\sum_{j=1}^{m} d\left(\bar{r}_j^k,\bar{r}^k\right)}\right)\right]}{\ln(n)} \tag{2.19}$$

$$\eta^k = 1 - e^k / \sum_{k=1}^{K}\left(1 - e^k\right) \tag{2.20}$$

where η^k is the stage weight, $\bar{r}^k$ and e^k are the intermediate parameters for the calculation.

Step 3: Rank the Order of the Components

On the basis of the comprehensive risk decision data of the whole period for the operational risk assessment, the final component risk ranking result on the basis of the maximum group utility Si, the minimum of individual regret Ri and the comprehensive risk value Qi can be acquired in accordance with Formulas (2.21)–(2.23).

$$S_i = \sum_{k=1}^{K}\sum_{j=1}^{m}\eta^k\omega_j\left(\frac{d\left(r_j^{k*}, r_{ij}^k\right)}{d\left(r_j^{k*}, r_j^{k-}\right)}\right) \tag{2.21}$$

$$R_i = \max_j\left[\eta^k\omega_j\left(\frac{d\left(r_j^{k*}, r_{ij}^k\right)}{d\left(r_j^{k*}, r_j^{k-}\right)}\right)\right] \tag{2.22}$$

$$Q_i = \nu\frac{(S_i - S^*)}{(S^- - S^*)} + (1-\nu)\frac{(R_i - R^*)}{(R^- - R^*)} \tag{2.23}$$

2.2.3 Case Study

In this paper, the operational risk assessment on the basis of TFNIFS and the dynamic VIKOR approach is validated by taking a specific railway train bogie system as an instance. Moreover, the bogie system has been applied to establish the multiplex network model and can be demonstrated in Fig. 2.5 [35].

On the basis of the railway train bogie system multiplex network model, the correlation risk index can be calculated with the fusions of the proposed dynamic properties in accordance with Formulas (2.1) and (2.2). Figure 2.6 illustrates the aforementioned properties with an example within the context of Stage 1, where the horizontal axis is the component number and the vertical one is the dynamic properties with failure of railway train bogie system. Furthermore, the correlation

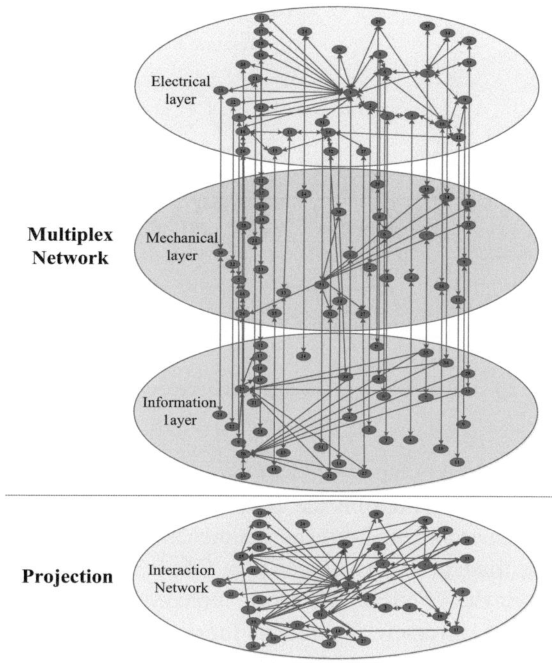

Fig. 2.5 Multiplex network model of railway train bogie system

risk index of these three constant stages can be systematically consolidated and depicted in Fig. 2.7.

As demonstrated in Fig. 2.7, the correlation risk index of each component is in the inter9val of 0.1 and 0.7, where the risk degree of the dynamic failure properties takes on an ascending trend with time. The frame assembly represents the highest correlation risk index.

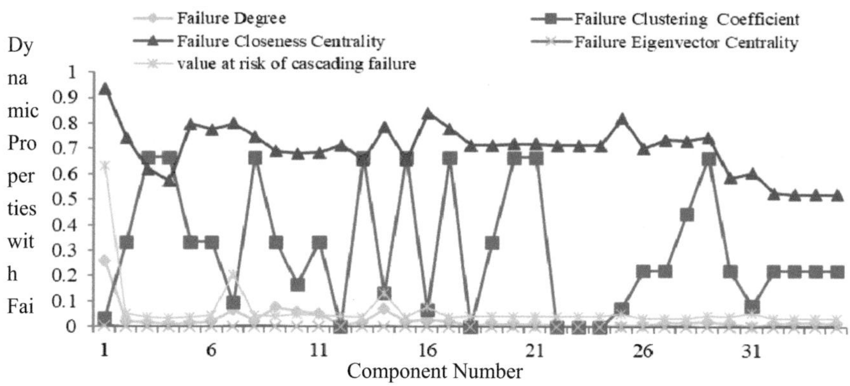

Fig. 2.6 Properties of railway train bogie system under stage 1

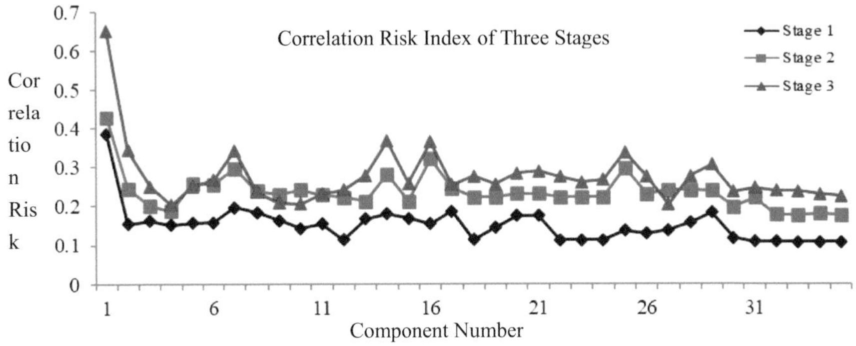

Fig. 2.7 Correlation risk index of three stages

Take the component "axle box body" (component 7) as an example, ((0.5, 0.6, 0.7), (0.2, 0.3, 0.4)) is the TFNIFS value of the risk effect index on system, people, and environment from the maintenance personnel D_1 in the assessment stage 2. The triangle fuzzy number (0.5, 0.6, 0.7) is characterized as a membership value within the TFNIFS framework, representing the positive degree associated with the index. Conversely, the triangle fuzzy number (0.2, 0.3, 0.4) is another fuzzy number denoting the non-membership within TFNIFS, depicting the negative degree related to the index. It is readily apparent that the positive degree significantly surpasses the negative degree, indicating the expert's perspective that the risk impact index on the system, individuals, and the environment is more likely to be higher and has an impact on the bogie system. Design manufacturer D_2 and railway train driver D_3 also give the evaluation result ((0.5, 0.6, 0.7), (0.2, 0.3, 0.4)) for the component "axle box body" (component 7) in the assessment stage 2.

Figure 2.8 provides an illustration of the risk effect index trend on the system, people, and environment for the "axle box body" component (Component 7) during three consecutive assessment stages, as reported by maintenance personnel D_1. The blue, pink, and red data points ((0.3, 0.4, 0.5), (0.4, 0.5, 0.6)), ((0.5, 0.6, 0.7), (0.2, 0.3, 0.4)), and ((0.7, 0.8, 0.9), (0.0, 0.1, 0.2)) represent the index values under Stage 1, Stage 2, and Stage 3, respectively. Clearly, the upward trajectory indicates a rising trend, signifying the expert's observation that the risk effect on the system, people, and environment is escalating over time. This trend underscores the notion that the occurrence and growth of risk in railway train components exhibit a nearly linear relationship with time.

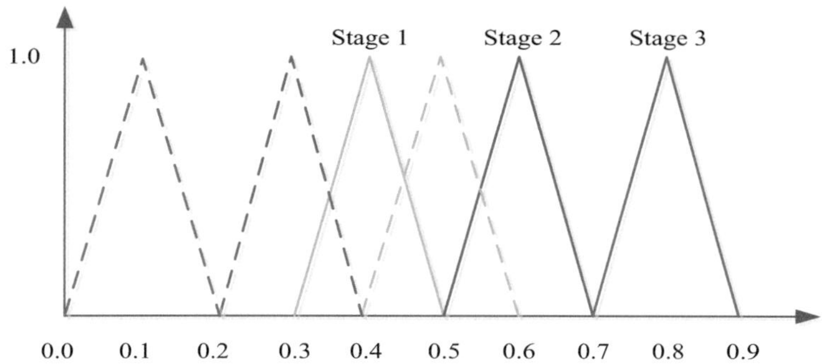

Fig. 2.8 Trend of risk effect index on system, people, and environment over time for axle box body

Subsequently, the qualitative assessment values of the risk effect index on the system, people, and environment, along with the fault detectability index and environment risk index, can be amalgamated into the comprehensive assessment value for each index. This integration is conducted in accordance with Formulas (2.4) and (2.5), taking into consideration expert credibility.

Table 2.4 provides an overview of the comprehensive value of the risk effect index on system, people, and environment on the basis of these three experts and engineers. Moreover, the weight of D_1, D_2, and D_3 under the distinct three stages can be determined as follows.

- $\lambda_1^1 = 0.43, \lambda_2^1 = 0.36, \lambda_3^1 = 0.21;$
- $\lambda_1^2 = 33, \lambda_2^2 = 0.35, \lambda_3^2 = 0.32;$
- $\lambda_1^3 = 34, \lambda_2^3 = 0.32, \lambda_3^3 = 0.34.$

The weighting of experts and engineers in one stage differs significantly from the other, a distinction attributed to the considerable disparity in their roles and working

Table 2.4 Comprehensive risk effect value on system, people, and environment for first 10 components

No.	Stage 1	Stage 2	Stage 3
1	((0.7,0.8,0.9), (0.0,0.1,0.2))	((0.5,0.6,0.7), (0.2,0.3,0.4))	((0.7,0.8,0.9), (0.0,0.1,0.2))
2	((0.7,0.8,0.9), (0.0,0.1,0.2))	((0.5,0.6,0.7), (0.2,0.3,0.4))	((0.7,0.8,0.9), (0.0,0.1,0.2))
3	((0.5,0.6,0.7), (0.2,0.3,0.4))	((0.5,0.6,0.7), (0.2,0.3,0.4))	((0.3,0.4,0.5), (0.4,0.5,0.6))
4	((0.3,0.4,0.5), (0.4,0.5,0.6))	((0.1,0.2,0.3), (0.6,0.7,0.8))	((0.5,0.6,0.7), (0.2,0.3,0.4))
5	((0.3,0.4,0.5), (0.4,0.5,0.6))	((0.1,0.2,0.3), (0.6,0.7,0.8))	((0.3,0.4,0.5), (0.4,0.5,0.6))
6	((0.3,0.4,0.5), (0.4,0.5,0.6))	((0.5,0.6,0.7), (0.2,0.3,0.4))	((0.3,0.5,0.7), (0.1,0.3,0.5))
7	((0.3,0.4,0.5), (0.4,0.5,0.6))	((0.5,0.6,0.7), (0.2,0.3,0.4))	((0.7,0.8,0.9), (0.0,0.1,0.2))
8	((0.1,0.2,0.3), (0.6,0.7,0.8))	((0.1,0.2,0.3), (0.6,0.7,0.8))	((0.0,0.0,0.1), (0.8,0.9,0.9)
9	((0.1,0.2,0.3), (0.6,0.7,0.8))	((0.3,0.4,0.5), (0.4,0.5,0.6))	((0.5,0.6,0.7), (0.2,0.3,0.4))
10	((0.5,0.6,0.7), (0.2,0.3,0.4))	((0.5,0.6,0.7), (0.2,0.3,0.4))	((0.7,0.8,0.9), (0.0,0.1,0.2))

environments. Accordingly, cognition and preference can make a large change to the risk matters with the bogie system. Additionally, the information obtained from the former stage can exert a significant influence on the assessment of subsequent stages. The information quantity and quality acquired from distinct stages can also vary significantly from each other. Consequently, the weight of each expert in distinct stages can have a great difference.

The mixed fuzzy decision matrix under distinct stages can be set up on the basis of the risk factors with MDBF, MTTR, risk effect on system, people and environment, maintenance cost, fault detectability, as well as correlation risk, operation skills, mental status, environment risk and TQI. Moreover, Table 2.5 illustrates a mixed fuzzy decision matrix under stage 2, which simply enumerates the first 10 numbers in the alternative components.

In the process of the dynamic VIKOR approach, r_j^{k*} and r_j^{k-} of the mixed fuzzy decision matrix can be determined to provide the basic preparation for the following criteria weight calculation under the distinct stages. Furthermore, the weight of each criterion under the distinct stage can be demonstrated as follows.

- $\omega^1 = [0.03\ 0.05\ 0.05\ 0.02\ 0.04\ 0.03\ 0.25\ 0.25\ 0.03\ 0.25]$;
- $\omega^2 = [0.03\ 0.02\ 0.04\ 0.03\ 0.02\ 0.03\ 0.27\ 0.27\ 0.02\ 0.27]$;
- $\omega^3 = [0.03\ 0.04\ 0.04\ 0.03\ 0.02\ 0.03\ 0.26\ 0.26\ 0.03\ 0.26]$.

Table 2.5 Mixed fuzzy decision matrix of stage 2

No.	C_1	C_2	C_3	C_4	C_5	C_6	C_7	C_8	C_9	C_{10}
1	0.25	0.3	((0.7,0.8,0.9), (0.0,0.1,0.2))	[0.5, 0.5]	((0.5,0.6,0.7), (0.2,0.3,0.4))	0.6	1.0	1.0	((0.5,0.6,0.7), (0.2,0.3,0.4))	1.0
2	0.11	0.5	((0.5,0.6,0.7), (0.2,0.3,0.4))	[1.0, 1.0]	((0.5,0.6,0.7), (0.2,0.3,0.4))	0.3	1.0	1.0	((0.3,0.4,0.5), (0.6,0.7,0.8))	1.0
3	0.61	0.4	((0.5,0.6,0.8), (0.2,0.3,0.4))	[1.0, 1.0]	((0.3,0.4,0.5), (0.6,0.7,0.8))	0.3	1.0	1.0	((0.5,0.6,0.8), (0.2,0.3,0.4))	1.0
4	0.08	0.5	((0.0,0.0,0.1), (0.8,0.9,0.9))	[0.71, 0.8]	((0.5,0.6,0.8), (0.2,0.3,0.4))	0.2	1.0	1.0	((0.3,0.4,0.5), (0.4,0.5,0.6))	1.0
5	0.23	0.3	((0.1,0.2,0.3), (0.6,0.7,0.8))	[1.0, 1.0]	((0.3,0.4,0.5), (0.6,0.7,0.8))	0.3	1.0	1.0	((0.3,0.4,0.5), (0.6,0.7,0.8))	1.0
6	0.35	0.5	((0.3,0.4,0.5), (0.6,0.7,0.8))	[0.8,0.83]	((0.1,0.2,0.3), (0.6,0.7,0.8))	0.3	1.0	1.0	((0.5,0.6,0.8), (0.2,0.3,0.4))	1.0
7	0.14	0.5	((0.5,0.7,0.9), (0.1,0.3,0.5))	[0.8, 0.83]	((0.5,0.6,0.8), (0.2,0.3,0.4))	0.3	1.0	1.0	((0.3,0.4,0.5), (0.6,0.7,0.8))	1.0
8	1.0	0.5	((0.1,0.2,0.3), (0.6,0.7,0.8))	[0.8, 0.83]	((0.3,0.4,0.5), (0.6,0.7,0.8))	0.2	1.0	1.0	((0.7,0.8,0.9), (0.0,0.1,0.2))	1.0
9	0.38	1.0	((0.3,0.4,0.5), (0.6,0.7,0.8))	[1.0, 1.0]	((0.7,0.8,0.9), (0.0,0.1,0.2))	0.3	1.0	1.0	((0.3,0.4,0.5), (0.6,0.7,0.8))	1.0
10	0.25	0.4	((0.5,0.6,0.8), (0.2,0.3,0.4))	[0.71, 0.8]	((0.7,0.8,0.9), (0.0,0.1,0.2))	0.3	1.0	1.0	((0.5,0.6,0.8), (0.2,0.3,0.4))	1.0

Then, the comprehensive data can be integrated on the basis of the weight of each stage, which can be determined by the Formulas (2.18) to (2.20). The weight ηk of each assessment stage can be demonstrated as.

- $\eta^1 = 0.24$;
- $\eta^2 = 0.37$;
- $\eta^3 = 0.39$.

In the end, the maximum group utility Si, the minimum of individual regret R_i, and the comprehensive risk value Q_i can be calculated in accordance with Formulas (2.21)–(2.23). Besides, Fig. 2.9 illustrates the results of Q_i, and indicates the ranking order of the alternatives. Furthermore, Fig. 2.9 also shows a comparison between the results with the static assessment [30]. From the result comparison between static and dynamic assessment, frame assembly is still the highest risk component. The risk degree of brake clamp, wheels, and axle box body exceeds the state of static assessment. In addition, these two distinct results make it clear that the relevant risk information and the expert cognition can experience constant and extreme change over time. Moreover, the dynamic operational risk assessment is conducted with multiple stages in accordance with the system and component maintenance. Except for that, the relevant risk information can be updated in this dynamic process, enhancing its reliability and effectiveness in evaluating the risk status of railway trains. Consequently, operational risk assessment on the basis of TFNIFS and dynamic VIKOR approach can provide technique support for railway train safety operation.

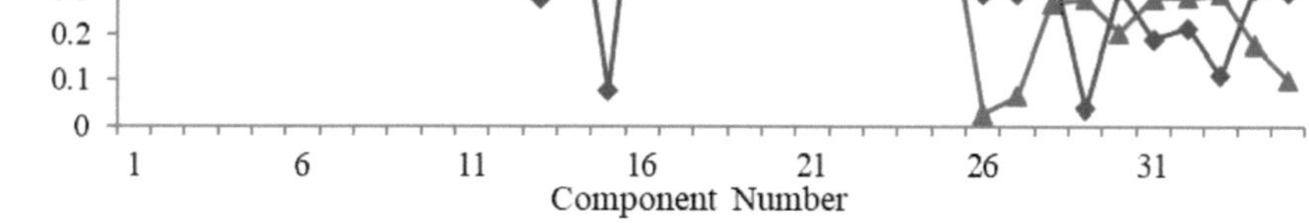

Fig. 2.9 Results comparison between static and dynamic assessment

2.3 The Dynamic Risk Analysis Method for High-Speed Train Operation Based on Computational Intelligence

2.3.1 *Dynamic Risk Analysis of High-Speed Train Operation on the Basis of Bayesian Networks*

2.3.1.1 Feature Variable Selection for Bayesian Network Models

The 15 input variables in the model, as mentioned in the research, include the causes of delays, severity of delay causes, and severity of delays. Regarding the causes of delays, to better comprehend the interrelationships among delay causes and their association with the delay phenomenon, expert consultations and pertinent guidelines from the railway authorities' fault recovery manuals were employed. The causes of delays were categorized into three categories: human factors, equipment factors, and external environmental factors. Among them, each category of delay factors includes several sub-nodes, and the specific summary of the initial delay factors affecting trains is indicated in Table 2.6.

Concerning the severity of delay causes, the goal of this case study is to examine the impact of fault severity on the extent of delay. Consequently, the grouping of fault recovery duration should be supervised by the variable "degree of delay." In this case, the selected approach is MDLP-based entropy grouping, which is on the basis of the Minimum Description Length Principle. Accordingly, the values for the severity of delay causes variable are demonstrated in Table 2.7.

Regarding the severity of delays, according to the regulations of the National Railway Group, a delay duration of 5 min or more is regarded as a delay. Consequently, when the delay time is less than 5 min, it is considered as no delay, and the variable value is set to 0. In the following, the data with a delay duration of 5 min or more is grouped. To ensure a balanced sample size for each group, the quantile grouping method is selected for data processing through SPSS Modeler software. This method is an unsupervised data grouping method, which implies that the grouping is based solely on the characteristics of the target object without considering the influence of other variables. Accordingly, the values for the severity of delays variable are demonstrated in Table 2.8.

2.3.1.2 Construction of Bayesian Network Model

Typically, on the condition that applying a Bayesian network model to solve practical engineering problems, the modeling process can be divided into four steps:

1. Abstract analysis of the actual engineering problem to determine the number of nodes and their values in the model.

Table 2.6 Factors influencing initial delays of high-speed trains

Interfering factors	Variable values	Variable description
Human factors		
(1) Staff operational errors	Occurrence (1), Non-occurrence (0)	
(2) Delays arising from passengers	Occurrence (1), Non-occurrence (0)	Including situations including passenger smoking triggering smoke alarms, accidental activation of emergency brake valve, sudden illness requiring immediate stopping for medical treatment, sudden increase in passenger flow, etc
(3) Personnel trespassing	Occurrence (1), Non-occurrence (0)	Intrusion of unauthorized personnel into the restricted area of the high-speed rail track
Equipment factors		
(4) Faulty maintenance equipment	Occurrence (1), Non-occurrence (0)	Track, switch, tunnel, bridge, and other infrastructure equipment failures
(5) Catenary failure	Occurrence (1), Non-occurrence (0)	Faults arising from overhead line abnormalities, including power outage (tripping) or other issues with the overhead line system
(6) Pantograph failure	Occurrence (1), Non-occurrence (0)	Including pantograph current collection quality
(7) Power supply equipment failure	Occurrence (1), Non-occurrence (0)	Failure of traction inverters and other power supply equipment, excluding pantographs and overhead lines
(8) Monitoring and surveillance equipment	Alarm/failure (1), Normal condition (0)	Alarm/failure of laser detection devices, TEDS, and other monitoring and surveillance equipment
(9) rolling stock failure	Occurrence (1), Non-occurrence (0)	Failure of braking equipment, water supply equipment, and other train equipment excluding doors, onboard control systems, swaying, or abnormal noises
(10) ATP and other train control system faults	Occurrence (1), Non-occurrence (0)	Failure of train control equipment, primarily ATP, balises, and other signaling devices
(11) Failure of the red tape and other signaling systems	Occurrence (1), Non-occurrence (0)	Malfunction of station entry signals, station exit signals, signals along the track, or indications lost at track switches in stations or track sections, resulting in the display of red lights indicating non-train occupancy
(12) Train swaying or unusual noises	Occurrence (1), Non-occurrence (0)	
(13) Failure of the dispatching system	Occurrence (1), Non-occurrence (0)	Failure of the CTC system

(continued)

Table 2.6 (continued)

Interfering factors	Variable values	Variable description
Environmental factors		
(14) Adverse weather conditions	Occurrence (1), Non-occurrence (0)	Including typhoons, thunderstorms, hailstorms, blizzards, etc
(15) Foreign object intrusion	Occurrence (1), Non-occurrence (0)	Intrusion of small animals or other objects into the boundary of the track

Table 2.7 Values of delay cause severity variable

Variable values	Handling methods:	Variable explanations
1	Field: "Event Duration" $\in (0, 34)$	Level I failure
2	Field: "Event Duration" $\in [34, 78)$	Level II failure
3	Field: "Event Duration" $\in [78, +\infty)$	Level III failure

Table 2.8 Values of delay severity variable

Variable values	Handling methods:	Variable explanations
0	Field: "Initial Delay Duration" $\in [0, 5)$	No delay
1	Field: "Initial Delay Duration" $\in [5, 10)$	Level I delay
2	Field: "Initial Delay Duration" $\in [10, 22)$	Level II delay
3	Field: "Initial Delay Duration" $\in [22, +\infty)$	Level III delay

2. Structural learning of the Bayesian network utilizing expert knowledge, domain expertise, or historical data, which involves using data learning algorithms or expert scoring methods to construct a directed acyclic graph.
3. Parameter estimation of the Bayesian network employing common methods including the Expectation–Maximization (EM) algorithm, which involves estimating the conditional probability tables for each node in the Bayesian network.
4. Leveraging the learning outcomes of the nodes, the Bayesian network's inference algorithm is employed to deduce the probability distribution of the states of specific nodes (Fig. 2.10).

There are several software options available for building Bayesian network models, including general-purpose software including MATLAB, R, and Python, as well as specialized software including Netica and GeNIe. Taking into account the requirements for model visualization and sensitivity analysis, GeNIe software is chosen for its proficiency in these domains.

(1) Bayesian Network Structure on the Basis of Model Assumptions

First, each type of delay factor is regarded as a parent node in the network, and the initial delay of the train is regarded as the child node of these parent nodes. The influence relationship between the delay factors and the initial delay is demonstrated

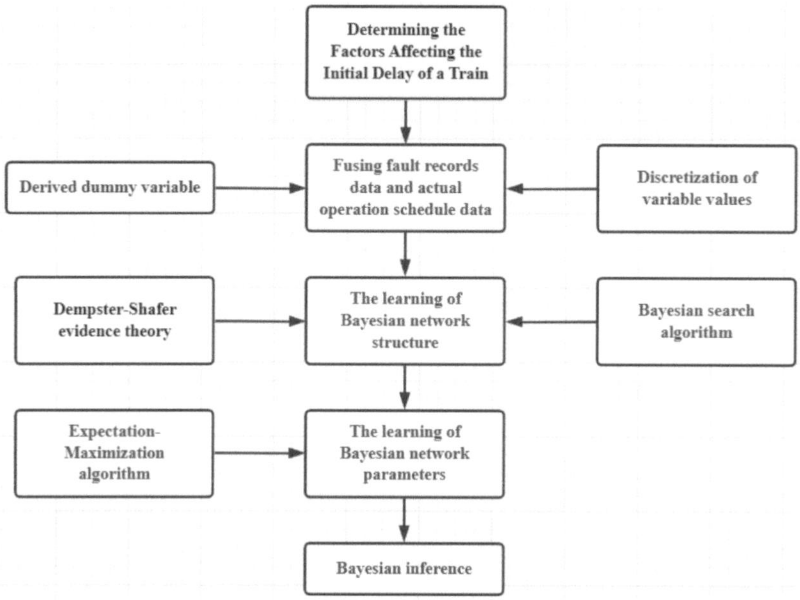

Fig. 2.10 Flowchart

by directed arcs. With an increase in the number of parent nodes, the conditional probabilities of the child node experience exponential growth, potentially leading to situations where some conditional probabilities cannot be assigned. Consequently, considering that the research has summarized a total of 15 types of delay factors, to attain an improved model fit, three intermediate nodes are added between the delay factor nodes and the delay severity parent node. These intermediary nodes are designated as personnel factors, equipment factors, and external environmental factors. The preliminary construction of the Bayesian network structure is demonstrated in Fig. 2.11.

(2) Network Structure on the Basis of Expert Experience

In the research, the expert scoring method was employed to obtain expert subjective experience. Besides, the scores provided by six experts on the relationship between delay influencing factors were integrated. Additionally, the D-S evidence fusion theory was employed to minimize the subjectivity of expert judgments and formulate a rational network structure.

Concerning expert selection, to enhance the reliability of subjective experience and minimize the impact of individual erroneous experiences on the data, questionnaires were distributed to six experts. Among them, two were on-site fault-handling personnel, and four were teachers and doctoral students specializing in railway safety research.

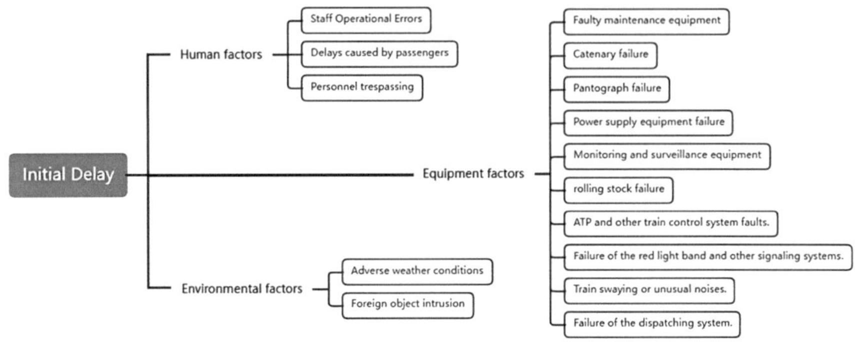

Fig. 2.11 Bayesian network structure on the basis of model assumptions

Concerning the design of the questionnaire, to mitigate the potential influence of ambiguous judgments on data accuracy, an additional category of relationship type was incorporated, namely, "uncertain relationship between variables." As a result, based on the conventional variable relationships, the variable relationships in this questionnaire were classified into the following four types:

1. The occurrence of variable A leads to the occurrence of variable B, denoted as $A \rightarrow B$.
2. The occurrence of variable B leads to the occurrence of variable A, denoted as $A \leftarrow B$.
3. Variable A and variable B are mutually independent, and there is no dependency between them, denoted as $A|B$.
4. The relationship between variable A and variable B is difficult to determine, denoted as $A \leftrightarrow B$.

When filling out the questionnaire, the experts need to assign weights to the four types of relationships between any two variables, and the sum of the weights assigned to the four types should be equal to 1 (Fig. 2.12).

In the data fusion process, the combination mass function between any two variables is calculated utilizing the evidence fusion theory. Moreover, the relationship type with the highest value in the mass function indicates the relationship type of the corresponding variable group.

The D-S evidence theory can effectively handle uncertain data and enhance the objectivity of the results. This method predominantly relies on conditional probabilities from Bayesian theory for calculations. Particularly, for the questionnaire data in this study, the calculation formula for the combination mass function is as follows:

$$M(A) = \frac{1}{K} \sum_{A_1 \cap A_2 \cap \cdots A_n = A} m_1(A_1) \cdot m_2(A_2) \cdots m_n(A_n) \tag{2.24}$$

Correlation among factors influencing the initial delay of high-speed rail trains.(Assignment can be to two decimal places.)								
Faulty maintenance equipment (I)	Adverse weather conditions (W)				Foreign object intrusion (F)			
	I→W	I←W	I↔W	I ∣ W	I→F	I←F	I↔F	I ∣ F
Assignment in [0,1]	0	0.8	0	0.2				
Catenary failure (N)	Adverse weather conditions (W)				Foreign object intrusion (F)			
	N→W	N←W	N↔W	N ∣ W	N→F	N←F	N↔F	N ∣ F
Assignment in [0,1]								
Pantograph failure (P)	Adverse weather conditions (W)				Foreign object intrusion (F)			
	P→W	P←W	P↔W	P ∣ W	P→F	P←F	P↔F	P ∣ F
Assignment in [0,1]								
Power supply equipment failure(E)	Adverse weather conditions (W)				Foreign object intrusion (F)			
	E→W	E←W	E↔W	E ∣ W	E→F	E←F	E↔F	E ∣ F
Assignment in [0,1]								
Monitoring & surveillance equipment(M)	Adverse weather conditions (W)				Foreign object intrusion (F)			
	M→W	M←W	M↔W	M ∣ W	M→F	M←F	M↔F	M ∣ F
Assignment in [0,1]								
Rolling stock failure (V)	Adverse weather conditions (W)				Foreign object intrusion (F)			
	V→W	V←W	V↔W	V ∣ W	V→F	V←F	V↔F	V ∣ F
Assignment in [0,1]								
Train door malfunction (D)	Adverse weather conditions (W)				Foreign object intrusion (F)			
	D→W	D←W	D↔W	D ∣ W	D→F	D←F	D↔F	D ∣ F
Assignment in [0,1]								
Train swaying or unusual noises (S)	Adverse weather conditions (W)				Foreign object intrusion (F)			
	S→W	S←W	S↔W	S ∣ W	S→F	S←F	S↔F	S ∣ F
Assignment in [0,1]								

Fig. 2.12 Example of questionnaire design

In formula,

$$K = \sum_{A_1 \cap \cdots \cap A_n \neq \phi} m_1(A_1) \cdot m_2(A_2) \cdots m_n(A_n)$$
$$= 1 - \sum_{A_1 \cap \cdots \cap A_n = \phi} m_1(A_1) \cdot m_2(A_2) \cdots m_n(A_n) \tag{2.25}$$

By utilizing expert scoring and the D-S evidence fusion theory, the causal relationships between variables were determined, as demonstrated in Fig. 2.13. The newly added causal relationships are illustrated by the blue-directed arcs. The six additional directed arcs clarify how alterations in the external environment can result in the occurrence of faults in specific equipment.

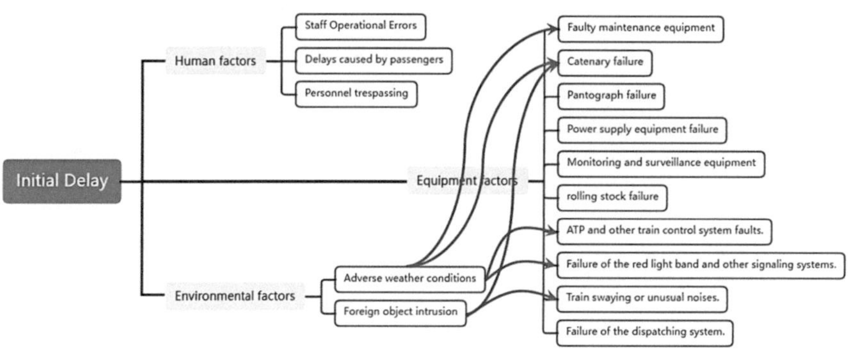

Fig. 2.13 Bayesian network structure based on expert experience

(3) Hybrid Network Structure on the Basis of Search Scoring Algorithm

The structure learning algorithms in GeNIe software can be categorized into two types: score-based and constraint-based algorithms. Due to the complete reliance on training samples for structure learning, the constraint-based algorithms are unsuitable for this scenario due to the presence of missing values in the data samples. Additionally, regarding a large number of training samples, the research utilizes the score-based algorithm termed GTT (Greedy thick thinning) for structure learning.

The scoring function of the GTT algorithm is on the basis of Bayesian principles and includes two types of scoring: K2 scoring and BDeu scoring. Both scoring metrics represent expansions of the Bayesian Dirichlet (BD) scoring function, with adjustments made to the allocation of Dirichlet parameters.

The K2 scoring function assumes the parameter to be a constant assigned with a value of 1. The scoring function is expressed as follows:

$$f_{K2}(G|D) = \sum_{i=1}^{n}\sum_{j=1}^{q_i}\left(\log\frac{(r_i-1)!}{(m_{ij}+r_i-1)!}\sum_{k=1}^{r_i}\log m_{ijk}!\right) + \log P(G) \tag{2.26}$$

Assuming $a_{ijk} = \frac{m}{r_i q_i}$, the BDeu scoring function is given by

$$\log P(G|D) = \sum_{i=1}^{n}\sum_{j=1}^{q_i}\left(\log\frac{\Gamma\left(\frac{m'}{q_i}\right)}{\Gamma\left(m_{ij}+\frac{m'}{q_i}\right)}\sum_{k=1}^{r_i}\log\frac{\Gamma\left(m_{ijk}+\frac{m'}{r_i q_i}\right)}{\Gamma\left(\frac{m'}{r_i q_i}\right)}\right) + \log P(G) \tag{2.27}$$

Concerning the search algorithm, GTT uses the GES (Greedy Equivalent Search) algorithm, which consists of two stages starting from an empty graph (no edges between nodes):

(1) GFS (Greedy Forward Search) stage: In this stage, a thick arc (directed edge) is added to the existing structure. Furthermore, the scoring function values of the original structure and the new structure are compared, and the structure with a superior score is preserved. This process is iterated until the score can no longer be improved.
(2) GBS (Greedy Backward Search) stage: In this stage, the obtained structure from the GFS stage undergoes edge deletion (thinning). After each removal of a directed edge, a comparison of the score is conducted between the new and old structures, and the structure with a higher score is retained. This process is iterated until further score improvement is not possible. At this juncture, the resulting structure represents the final Bayesian network.

In this case, the chosen algorithm for GTT is on the basis of the K2 scoring function. Moreover, the network structure includes 15 nodes indicating the occurrence of delay causes and the initial delay level of trains. Furthermore, in order to amalgamate the network structure acquired through data learning with the model derived from

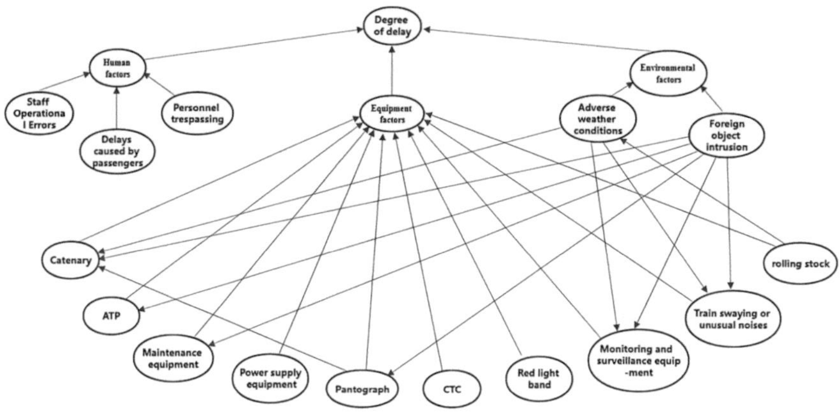

Fig. 2.14 Bayesian network structure on the basis of expert experience and data learning corrections

expert experience, background knowledge encompassing the expert-based network structure is incorporated to guide the structure learning from data.

The network structure obtained from GTT algorithm learning is demonstrated in Fig. 2.14. In comparison to the network structure learned from expert experience, the new network structure includes 5 additional directed arcs.

(3) Parameter Estimation on the Basis of the EM Algorithm.

Due to the presence of incomplete records in the incidents of sudden faults during train operation, including the event record of pantograph failure arising from power supply equipment, which is denoted as "automatic lowering of pantograph", and taking into account the restricted availability of training samples for certain factors contributing to delays, there are missing data. Consequently, the EM algorithm is selected to compute the parameters of the Bayesian network.

The EM algorithm is an iterative hill-climbing algorithm that consists of two steps: the E-step (Expectation) and the M-step (Maximization). By iteratively performing the E-step and M-step until convergence, the algorithm estimates the unknown parameters. The basic framework of the EM algorithm is demonstrated in Fig. 2.15.

2.3.1.3 Results Analysis

The computational results of the model are demonstrated in Fig. 2.16, where the bar charts for each node display the corresponding variables' marginal probabilities. Upon scrutinizing the bar charts, it becomes evident that contact wire faults and foreign object intrusion are the most prevalent types of faults, succeeded by red tape faults, shaking or noise, and personnel intrusion.

Bayesian inference can be categorized into two primary types: predictive inference and diagnostic inference. Moreover, predictive inference aims to predict the possible

The Basic Framework of the EM Algorithm
1. Input: the structure of a Bayesian network, training set data Đ, convergence threshold δ.
2. Assign initial values: $t = 0, \theta^0$ is assigned a random value.
3. The current score is $l(\theta^t \mid Đ)$.
4. while(true)
5. Step E: compute $m_{ijk}^t (i = 1,2, \dots, n; j = 1,2, \dots, q_i; k = 1,2, \dots, r_i)$
6. Step M: compute θ^{t+1}
7. The updated score is $l(\theta^{t+1} \mid Đ)$.
8. if(The updated score> The current score+δ)
9. Update the current score to $l(\theta^{t+1} \mid Đ)$.
10. t = t+1
11. else
12. return θ^{t+1}
13. end if
14. end while
15. Output: parameter estimation in Bayesian networks.

Fig. 2.15 Basic framework of the EM algorithm

states of the train's initial delay on the basis of the distribution of the parent nodes' delay-causing factors, which aids the personnel in implementing emergency response plans at distinct levels in response to distinct fault events, thereby efficiently utilizing limited emergency resources for delay prevention and control. Conversely, diagnostic inference operates in an inverse manner. Its objective is to deduce the most probable states of the factors causing delays when presented with a particular delay state of the train. This aids personnel in formulating pertinent maintenance and inspection plans to mitigate the recurrence of such delay states, ultimately diminishing the frequency of delays.

(1) Predictive Inference

In predictive inference, the variable state with the highest posterior probability is regarded as the prediction result. In the context of the train delay model, where numerous delay-causing nodes with a considerable number of potential values exist, depending solely on manual calculations can be time-consuming, labor-intensive, and susceptible to errors. Consequently, for subsequent Bayesian inference tasks, the use of precise inference algorithms is preferred. In this scenario, the selected algorithm is the Junction Tree algorithm, commonly employed in GeNIe software. The Junction Tree algorithm utilizes the conditional independence properties inherent in the Bayesian network graph. It decomposes the intricate Bayesian network into a collection of smaller cliques, subsequently transforming them into a joint tree structure through specific methods. This allows for efficient inference and solving of probabilistic queries utilizing the properties and operations of the tree structure. By employing the Junction Tree algorithm, the computational complexity and accuracy of the predictive inference process can be substantially enhanced.

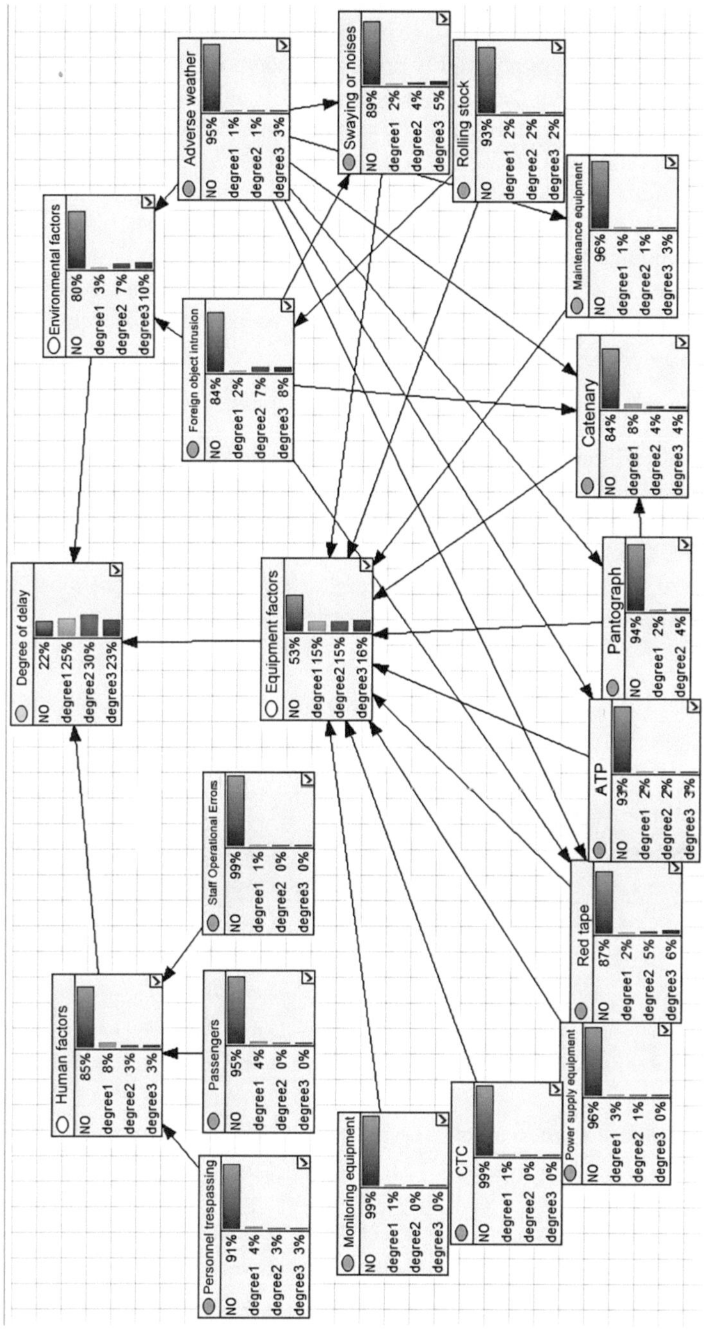

Fig. 2.16 Bayesian network probability distribution on initial delays of high-speed trains

By employing the Junction Tree algorithm, predictive inference is performed on the remaining 14 delay-causing nodes, and the prediction results are summarized in Table 2.9. In general, when the delay-causing factor is ascribed to human factors or environmental factors, the train is highly prone to encountering Level I and Level II delays. On the condition that the delay-causing factor is attributed to human factors with a short duration, the train is not likely to encounter delays, and its impact on normal train operation is minimal. On the condition that the delay-causing factor is attributed to equipment factors, the severity of the train delay is highly likely to be positively linked to the severity of the equipment failure. Nonetheless, the case of dispatching equipment failure is more specific. A brief duration of failure recovery is highly likely not to impact normal train operation. However, when the duration of the failure exceeds 34 min, the train is highly likely to experience Level II delays, characterized by a delay time surpassing 10 min.

(2) Diagnostic Inference

Set all root nodes of the delay causes as target nodes and select "Level I Failure," "Level II Failure," and "Level III Failure" as target states. Designate the child node "Delay Severity" as an observation node and designate all intermediate nodes as auxiliary nodes. Use the junction tree algorithm to perform diagnostic inference on the four states of the "Delay Severity" node and observe the posterior probabilities of the target nodes under distinct states of the observation node.

Table 2.9 Predicted delay severity of train delays due to distinct risk levels of delay causes

Fault type	Fault severity		
	Level I Fault	Level II Fault	Level III Fault
Staff operational error	No Delay	Level I Delay	Level II Delay
Passenger-induced delay	No Delay	Level I Delay	Level II Delay
Personnel trespassing	No Delay	Level I Delay	Level II Delay
Faulty maintenance equipment	Level I Delay	Level II Delay	Level III Delay
Catenary failure	Level I Delay	Level II Delay	Level III Delay
Pantograph failure	Level I Delay	Level II Delay	–
Power supply equipment failure	Level I Delay	Level II Delay	Level III Delay
Monitoring and surveillance equipment	Level I Delay	Level II Delay	Level III Delay
Rolling stock failure	Level I Delay	Level II Delay	Level III Delay
ATP and other train control system faults	Level I Delay	Level II Delay	Level III Delay
Failure of the red tape and other signaling systems	Level I Delay	Level II Delay	Level III Delay
Train swaying or unusual noises	Level I Delay	Level II Delay	Level III Delay
Failure of the dispatching system	No Delay	Level II Delay	Level II Delay
Adverse weather conditions	Level I Delay	Level II Delay	Level II Delay
Foreign object intrusion	Level I Delay	Level II Delay	Level II Delay

Based on Table 2.10, the posterior probability of a Level I Overhead Line Failure is the highest, increased in comparison to the prior probability of 0.081. By sorting the posterior probabilities of the target nodes in ascending order, it is obvious that the most probable factor influencing Level I delays is a Level I Overhead Line Failure, followed by Level III and Level II Foreign Object Intrusion, and Level II and Level III Red Light Strip Faults.

Based on Table 2.11, the posterior probability of a Level II Foreign Object Intrusion is the highest, increased in comparison to the prior probability of 0.069. Upon arranging the posterior probabilities of the target nodes in ascending order, it becomes apparent that the most probable factor influencing Level II delays is a Level II or Level III Foreign Object Intrusion, succeeded by Level III and Level II Red Light Strip Faults.

According to Table 2.12, the posterior probability of a Level III Foreign Object Intrusion is the highest, increased in comparison to the prior probability of 0.078. By arranging the posterior probabilities of the target nodes in ascending order, it is evident that the most likely factor influencing Level III delays is a Level III Foreign Object Intrusion, followed by Level III Red Light Strip Fault, Level III Shaking or Abnormal Noise, and Level III Contact Network Fault.

According to Table 2.13, the posterior probability of a Level I Contact Network Fault is the highest, increased compared to the prior probability of 0.081. By arranging the posterior probabilities of the target nodes in ascending order, it is evident that the least influential factors on normal train operation are Level I Contact Network Fault,

Table 2.10 Diagnosis results table for train delays with "Level I Delay" severity

Root node	Initial train delay = Level I Delay		
	Level I Fault	Level II Fault	Level III Fault
Staff operational error	0.010	0.003	0.001
Passenger-induced delay	0.037	0.006	0.001
Personnel trespassing	0.030	0.043	0.037
Faulty maintenance equipment	0.009	0.007	0.025
Catenary failure	0.117	0.036	0.031
Pantograph failure	0.029	0.040	–
Power supply equipment failure	0.038	0.007	0.003
Monitoring and surveillance equipment	0.009	0.003	0.004
Rolling stock failure	0.033	0.022	0.016
ATP and other train control system faults	0.027	0.018	0.023
Failure of the red tape and other signaling systems	0.020	0.047	0.047
Train swaying or unusual noises	0.022	0.041	0.040
Failure of the dispatching system	0.007	0.004	0.004
Adverse weather conditions	0.020	0.005	0.021
Foreign object intrusion	0.025	0.050	0.061

Table 2.11 Diagnosis results table for train delays with "Level II Delay" severity

Root node	Initial train delay = Level II Delay		
	Level I Fault	Level II Fault	Level III Fault
Staff operational error	0.007	0.002	0.001
Passenger-induced delay	0.025	0.003	0.001
Personnel trespassing	0.020	0.020	0.035
Faulty maintenance equipment	0.005	0.008	0.032
Catenary failure	0.049	0.049	0.047
Pantograph failure	0.014	0.054	–
Power supply equipment failure	0.020	0.008	0.003
Monitoring and surveillance equipment	0.005	0.004	0.004
Rolling stock failure	0.016	0.028	0.022
ATP and other train control system faults	0.015	0.023	0.030
Failure of the red tape and other signaling systems	0.011	0.063	0.071
Train swaying or unusual noises	0.011	0.055	0.058
Failure of the dispatching system	0.004	0.005	0.004
Adverse weather conditions	0.008	0.006	0.030
Foreign object intrusion	0.008	0.113	0.095

Table 2.12 Diagnosis results table for train delays with "Level III Delay" severity

Root node	Initial train delay = Level III Delay		
	Level I Fault	Level II Fault	Level III Fault
Staff operational error	0.008	0.002	<0.001
Passenger-induced delay	0.030	0.003	<0.001
Personnel trespassing	0.024	0.018	0.022
Faulty maintenance equipment	0.005	0.006	0.043
Catenary failure	0.046	0.034	0.067
Pantograph failure	0.014	0.038	–
Power supply equipment failure	0.021	0.007	0.004
Monitoring and surveillance equipment	0.005	0.003	0.006
Rolling stock failure	0.016	0.020	0.029
ATP and other train control system faults	0.015	0.017	0.040
Failure of the red tape and other signaling systems	0.010	0.043	0.099
Train swaying or unusual noises	0.011	0.038	0.079
Failure of the dispatching system	0.004	0.004	0.004
Adverse weather conditions	0.008	0.005	0.035
Foreign object intrusion	0.009	0.051	0.109

followed by Level I Passenger Factors and Level I Trespassing. Contact network faults, despite their frequent occurrence, frequently stem from events that can be rectified swiftly, leading to minimal impact on normal train operation. Delays arising from passengers are typically due to sudden illnesses requiring immediate medical attention, and yet the duration of stopping for the sick passenger is short and can be coordinated with passenger operations, and consequently minimizing train delays. Level I Trespassing incidents mostly involve unauthorized individuals entering restricted areas. The deployment of surveillance cameras along the railway lines for real-time monitoring enables railway personnel to swiftly identify and address trespassing situations. Additionally, individuals with normal awareness are unlikely to risk their lives by crossing the railway tracks when a train is approaching, further diminishing the impact on train punctuality.

In general, the most probable reasons for initial train delays include foreign object intrusion, contact network faults, red tape faults, and shaking or abnormal noise. Delays arising from foreign object intrusion and red tape faults are frequently more than 10 min. Additionally, for Level I delays, other substantial causes include locomotive and rolling stock faults, whereas for Level III delays, there is a higher probability of delays arising from ATP and other train control equipment faults.

Analyzing the aforementioned inference results can offer a scientific foundation for devising specialized detection and maintenance plans for various types of faults. This, in turn, aids in minimizing the cost of detection and maintenance while ensuring

Table 2.13 Diagnosis results table for no delay in train departure

Root node	Initial train delay = No delay		
	Level I Fault	Level II Fault	Level III Fault
Staff operational error	0.022	0.002	0.001
Passenger-induced delay	0.092	0.004	0.001
Personnel trespassing	0.077	0.026	0.023
Faulty maintenance equipment	0.009	0.006	0.022
Catenary failure	0.120	0.026	0.025
Pantograph failure	0.029	0.030	–
Power supply equipment failure	0.039	0.006	0.002
Monitoring and surveillance equipment	0.009	0.003	0.003
Rolling stock failure	0.033	0.016	0.013
ATP and other train control system faults	0.028	0.015	0.020
Failure of the red tape and other signaling systems	0.021	0.033	0.037
Train swaying or unusual noises	0.022	0.030	0.032
Failure of the dispatching system	0.007	0.003	0.004
Adverse weather conditions	0.022	0.005	0.015
Foreign object intrusion	0.027	0.049	0.039

the smooth operation of the railway system. In the daily risk management work, the following recommendations can be considered for optimization:

(1) Give high priority to fault types with high frequency and high delay rates, including shaking or abnormal noise, contact network faults, and red tape faults. Enhancing the frequency of inspections and maintenance for these fault types during routine operations. Rationalize and efficiently allocate detection and maintenance resources by considering the historical distribution, duration, and severity of faults.
(2) Give importance to the development of emergency response plans for low-frequency but high-delay-rate faults, including track maintenance equipment, signaling equipment, and pantograph faults. Despite the fact that the occurrence probability of these equipment faults is low, once they occur, the duration of the faults is typically more than 34 min, resulting in prolonged initial delays. In addressing this category of faults, it is crucial to concentrate on specific high-frequency fault events, such as the automatic lowering of the pantograph in the case of pantograph faults, and conduct comprehensive research on their spatiotemporal patterns and underlying causes. Optimizing the choice and distribution of maintenance resources, reducing on-site repair duration, responding to faults promptly, and minimizing the repercussions on train safety and operations.
(3) Do not underestimate high-frequency yet low-delay-rate fault types, including trespassing incidents. Despite the fact that they generally cause lower levels of delay, they pose remarkable safety hazards. Consequently, for this type of fault, a combination of "hardware and software" measures can be implemented.

Equipment and facilities: Constructing protective barriers in sections susceptible to trespassing incidents is aimed at preventing unauthorized entry. Augmenting video surveillance in areas challenging to isolate from the external environment, such as around overhead lines, is intended to facilitate comprehensive real-time monitoring, ensuring the prompt detection of trespassing incidents.

Rules and regulations: Implementing strict constraints on the work behavior of railway staff, standardizing and refining work processes, and conducting regular safety knowledge assessments. Providing public education on railway production safety to external individuals and strengthening penalties for unauthorized trespassing behaviors.

2.3.2 Prediction of High-Speed Train Emergency Incident Harm on the Basis of Random Forest

The number of delayed trains arising from emergency incidents is an important evaluation index to measure the impact of such incidents. Following an occurrence, if it impacts a greater number of trains, it signifies that the event's scope is broader, and the challenge of restoring normal train operation is heightened. This also indicates that the harm of the event is greater and demands increased focus on preventive

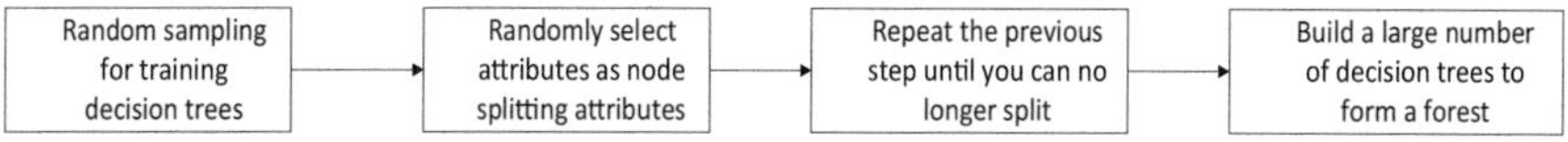

Fig. 2.17 Basic steps of random forest

measures. To further calculate the impact of distinct emergency incidents on the number of delayed trains and quantitatively analyze the severity of the emergency incidents, this section will establish a prediction model for the harm of high-speed train emergency incidents on the basis of random forest.

2.3.2.1 Overview of Random Forest

Random forest is a machine learning algorithm based on bootstrap aggregation. It was proposed by scholars Breiman and Cutler on the basis of the enhancement of the decision tree algorithm, and is a typical parallel ensemble learning algorithm. Moreover, the core of random forest is to build several regression decision trees on the basis of specified features, and subsequently combine them to form a forest. Ultimately, the outcomes of the decision trees within the forest are acquired through voting, thereby enhancing prediction accuracy without augmenting computational complexity. The fundamental steps are illustrated in Fig. 2.17.

In comparison with common machine learning algorithms, the random forest algorithm possesses the following advantages:

(1) Simple implementation, enhanced training speed through parallel computing, less prone to overfitting, as well as high accuracy;
(2) The algorithm has low requirements for data types and scales, and can readily adaptable to both continuous and discrete data;
(3) It has the capability to conduct classification, clustering, and regression analyses on high-dimensional data without necessitating dimensionality reduction or feature selection.
(4) It can assess the significance and correlation of feature variables.

2.3.2.2 Modeling Steps

On the basis of the analysis, the type, duration, location, and time of occurrence of an unexpected event will all influence the quantity of trains experiencing delays. For instance, the extended duration of the unforeseen event correlates with an increased impact on trains, consequently leading to a higher number of delayed trains. Additionally, if the unexpected event occurs during peak hours, when there is a high density of trains and little redundancy time in the train schedule, the delay experienced by one train due to interference has the potential to rapidly propagate to neighboring trains, ultimately resulting in widespread train delays.

Consequently, in the model for predicting the hazard level of unexpected events, the type, duration, location, as well as time of occurrence of the event are employed as the feature variables, and the number of delayed trains arising from the event is utilized as the target variable. An increase in the number of delayed trains corresponds to a heightened level of hazard associated with the event. By inputting the spatiotemporal distribution characteristics of the unexpected event, the number of delayed trains arising from the event can be obtained, and consequently the hazard level of the event can be determined (Table 2.14).

(1) Preparation

Import necessary libraries including pandas, numpy, sklearn, and matplotlib in Python 3.5.2 (Anaconda 3.2.0) environment.

(2) Data Preprocessing

Importing the organized dataset of 3422 train delay records without any missing data. It is clear that the event type and location are discrete features, and no quantitative relationship exists between them. Consequently, One-Hot encoding is chosen to transform them into numerical features, as demonstrated in Fig. 2.18. One-Hot encoding can transform a feature variable with n possible values into n mutually exclusive feature variables, and only one can be activated at a time. It not only broadens the scope of feature variables but also enhances the rationale behind distance and similarity calculations for discrete feature variables.

Table 2.14 Variable description

Variable	Description
Feature variable	Event types (vehicle failures, on-board equipment failures, track failures, power supply equipment failures, communication signal equipment failures, natural environmental interference, human interference, and dispatch monitoring equipment failures)
	Duration
	Location of occurrence (station/section)
	Time of occurrence (0–24 h)
Target variable	The number of delayed trains

	Duration	Occurrence time	Number of late trains	Event type_Human interference	Event type_Power supply device is faulty	Event type_Line fault	Event type_Natural environmental disturbance	Event type_Scheduling Monitoring device fault	Event type_On-board equipment fault	Event type_Vehicle failure	Event type_Communication signal device fault	Occurrence location_interval	Occurrence location_station
0	101	0.33936	1	0	1	0	0	0	0	0	0	1	0
1	26	0.416551	1	0	0	1	0	0	0	0	0	0	1
2	55	0.855355	1	0	0	0	0	0	1	0	0	0	1
3	122	0.403338	4	0	1	0	0	0	0	0	0	1	0
4	109	0.424896	48	0	0	0	1	0	0	0	0	1	0
5	25	0.631433	4	0	0	0	0	0	0	1	0	0	1
6	42	0.445063	8	0	0	0	0	0	0	1	0	1	0
7	49	0.739917	1	0	0	1	0	0	0	0	0	0	1
8	41	0.654381	8	0	1	0	0	0	0	0	0	1	0
9	26	0.557719	1	0	0	1	0	0	0	0	0	0	1

Fig. 2.18 One-hot encoding representation

In addition, the occurrence time of the emergency event is from 0:00 to 24:00. To enhance the processing speed and solution accuracy of the data, the time is converted to seconds and subsequently normalized utilizing min–max normalization, as demonstrated in formula 2.34, to scale 0:00–24:00 to the [0,1] interval.

$$x' = \frac{x - x_{\min}}{x_{\max} - x_{\min}} \tag{2.28}$$

where x is the original value, $x_{\min}$ and $x_{\max}$ indicate the minimum and maximum values in the original value, and x' is the normalized value.

(3) Building a Random Forest Regression Model

(1) Separating the Feature Variables and Target Variable: Following data preprocessing, the feature variables in the model are expanded to 12, including duration, occurrence time, event type_human interference, event type_power supply equipment failure, event type_line failure, event type_natural environment interference, event type_dispatch monitoring equipment failure, event type_on-board equipment failure, event type_vehicle failure, event type_communication signal equipment failure, occurrence location_interval, occurrence location_station, and the target variable is the number of delayed trains.
(2) Splitting the Dataset: The 3422 data points are partitioned into training and testing sets, where the initial 2567 data points constitute the training set, and the remaining 855 data points form the testing set.
(3) Building the Model: Initially, the RandomForestRegressor() model is invoked. Subsequently, the RandomizedSearchCV function and GridSearchCV function are used to adjust the model parameters. Besides, the former method randomly selects parameter combinations within a specified range for a predetermined number of iterations to attain the optimal parameter combination, whereas the latter iterates through all parameter combinations to obtain the optimal parameter combination. Typically, the two functions are integrated, employing the first function to acquire the approximate range of optimal parameters, while the second function is applied for refining the parameters within a specific range. The optimal parameter combination is obtained as demonstrated in Fig. 2.19, and the random forest model is built and trained on the basis of the optimal parameter combination.

Fig. 2.19 Optimal parameter combination

```
{'bootstrap': True,
 'max_depth': 10,
 'max_features': 3,
 'min_samples_leaf': 4,
 'min_samples_split': 10,
 'n_estimators': 100}
```

(4) Model Evaluation

Following training the model, relevant metrics are required to assess the efficacy of regression predictions. In this case, the evaluation metrics employed are the coefficient of determination (R2), explained variance score (EVS), mean squared error (MSE), and mean absolute error (MAE). The initial two metrics possess a value range of [0,1], wherein values closer to 1 signify enhanced explanatory power of the independent variable concerning variations in the dependent variable, indicating a better fit for the model. The latter two metrics reflect the degree of closeness between predicted and actual data, with smaller values signifying improved fit.

(5) Calculation of Feature Variable Importance

Through the random forest model, the influence of all feature variables on the target variable can be obtained, which is thess feature variable importance. Besides, there are two common calculation methods: Gini index method and precision reduction method. Here, in the context, the first method is selected, and for the feature variable m, the sum of the Gini index reduction degree of the branch nodes formed by m in each tree is the importance of the feature variable m.

(6) Prediction of Late Train Number on the Basis of Model Input Feature Variables (Fig. 2.20).

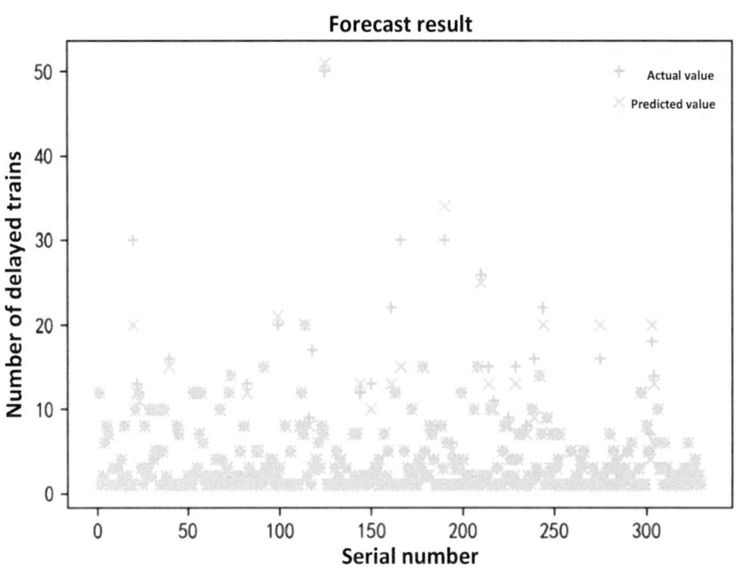

Fig. 2.20 Random forest regression prediction results for the number of delayed trains

2.3.2.3 Model Analysis

(1) Model Evaluation Table

Table 2.15 demonstrates the evaluation results of the prediction model. As evident from the table, the goodness of fit for the training set and test set of the predictive model concerning the severity of sudden events is 0.9801 and 0.9774, respectively. This suggests that the model exhibits commendable performance and is suitable for analyzing the severity of sudden events.

330 records were randomly selected from the sudden event dataset and validated by inputting them into the prediction model. Figure 2.18 indicates the prediction results of the model. It is obvious that the predicted values of the model are largely in accordance with the actual values, indicating that the model demonstrates strong predictive performance.

(2) Analysis of Feature Variable Importance

Following calculation, the importance of all feature variables is demonstrated in Fig. 2.21 and Table 2.16. By analyzing importance, the focal points of daily prevention and emergency management tasks can be elucidated.

From the results, it is evident that the two most crucial feature variables are duration (0.47) and occurrence time (0.25), indicating that these two variables have the highest impact on train delays. In practice, irrespective of the nature of the unforeseen event, its occurrence during peak rush hours or prolonged duration tends to adversely affect the regular operation of a significant number of trains. Considering that both duration and occurrence time are time-based features of unexpected events, it illustrates that the distribution of time is a pivotal factor in assessing the degree of hazard. The next important variables are natural environmental interference (0.07), sections (0.06), and stations (0.05).

Natural environmental interference belongs to external interference. Upon the onset of extreme weather or geological disasters, there will be a detrimental impact on all trains currently in operation or scheduled to depart on the affected lines. It will also affect the normal operation of other equipment and facilities within the high-speed railway system, leading to a large number of train delays. Sections and stations belong to spatial features. Moreover, the location where the unexpected event occurs will affect the emergency response methods for subsequent events. For instance, in the event of a train bogie failure, if it transpires at a station, maintenance personnel can be promptly contacted to undertake repairs or replace the train undercarriage.

Table 2.15 Evaluation metrics for random forests

Evaluation metric	Training set	Test set
R2	0.9801	0.9774
EVS	0.9788	0.9756
MSE	0.5142	0.5103
MAE	0.1399	0.2299

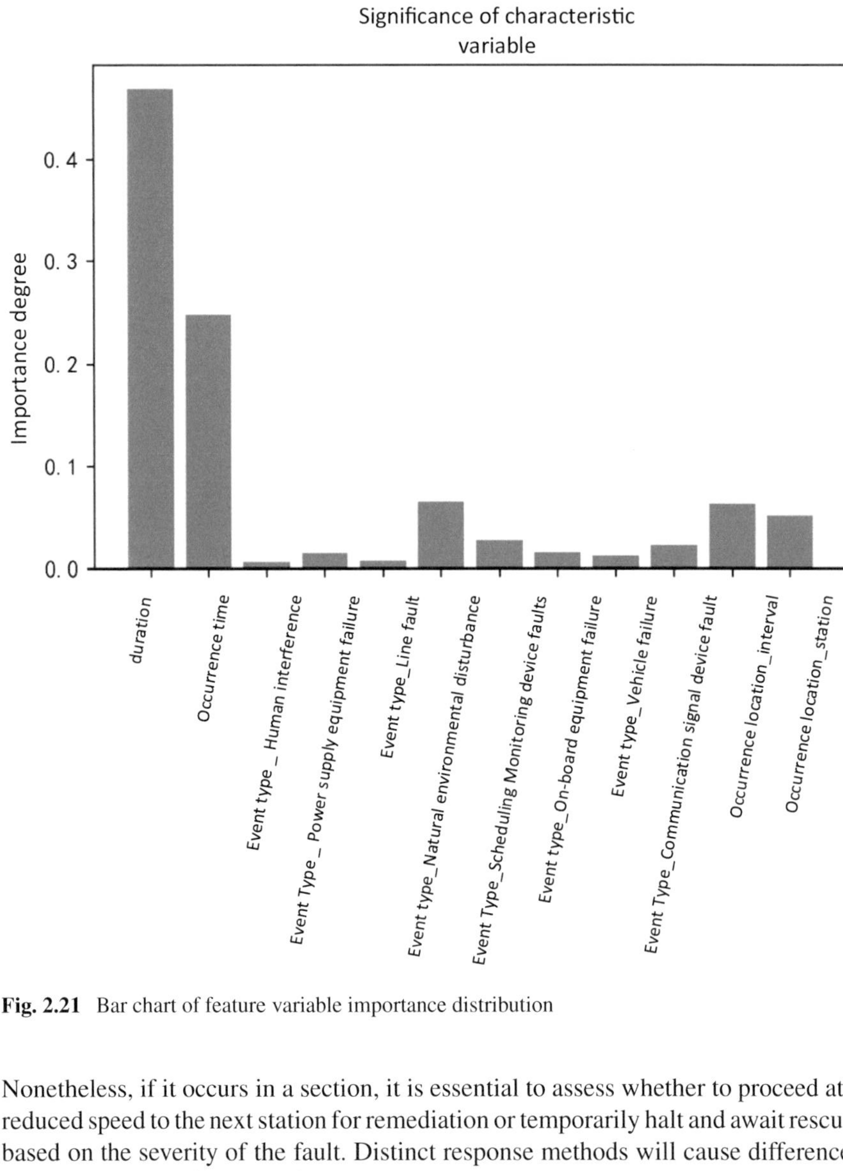

Fig. 2.21 Bar chart of feature variable importance distribution

Nonetheless, if it occurs in a section, it is essential to assess whether to proceed at a reduced speed to the next station for remediation or temporarily halt and await rescue, based on the severity of the fault. Distinct response methods will cause differences in the number of affected trains.

The types of unexpected events are slightly less essential. According to the previous analysis, regardless of the type of unforeseen event, the occurrence time, location, and duration are variable. Consequently, merely knowing the type of event is insufficient for accurately assessing the degree of hazard. For event types, the importance of three types of events is higher, namely, dispatch monitoring equipment failure, onboard equipment failure, as well as communication signal equipment failure, while the importance of three types of events, including human interference,

Table 2.16 Ranking of feature variable importance

Feature variables	Variable ranking
Event type	0.489
Duration	0.244
Location	0.121
Duration	0.47
Time of occurrence	0.25
Natural environmental interference	0.07
Interval	0.06
Station	0.05
Dispatch monitoring equipment failure	0.03
Onboard equipment failure	0.02
Communication signal equipment failure	0.02
Human interference	0.01
Power supply equipment failure	0.01
Line failure	0.01
Vehicle failure	0.01

power supply equipment failure, and line failure, and vehicle failure, is lower. It is evident that, despite the infrequent occurrence of dispatch monitoring equipment failures, when such issues arise, dispatchers lose the capability to monitor train statuses in real-time and issue dispatch commands. This loss of remote control over multiple trains has a significant impact on train delays. Despite the fact that power supply equipment failures occur frequently, their duration is relatively short, resulting in a smaller impact on train delays.

On the basis of the predicted model of the hazard level of unexpected events on delayed trains, understanding the potential impact of the event on regular train operations allows for the formulation of an appropriate operational adjustment plan. This plan, based on the number of affected trains, ensures the successful completion of transportation services under normal conditions.

2.4 Summary

The operational risk of railway train represented by high-speed trains includes the comprehensive risk emphasizing system failure and operational risk focusing on delay events. This chapter initially focuses on the all-encompassing risk associated with the railway train systems. It introduces a comprehensive risk analysis method for train systems based on a Type II intuitionistic fuzzy set and the dynamic VIKOR method. The approach considers the performance risk of the train system, taking into account the multi-stage and multi-factor characteristics of external condition risks, and aligns with the practicalities of the maintenance plan. Meanwhile, this method can analyze the comprehensive risk of train system and identify the key components with high risk. Secondly, this chapter employed the Bayesian network, which can more effectively capture the interdependence among variables, to study the correlation between the initial delay of high-speed trains and the causes of delay. Characteristic variables were identified, and 371 emergencies were categorized into 15 types of delay causes. In the network construction, the D-S evidence fusion theory was used to process the questionnaire data to acquire the initial structure, the GTT (Greedy thick thinking) algorithm was utilized to optimize the structure, and expert experience and data learning algorithms were employed to learn the structure of Bayesian network model. EM algorithm was adopted to learn model parameters. Besides, the learned Bayesian network model was used for reasoning research, and the initial delay state of trains in the event of emergencies with distinct fault levels was predicted. The types of emergencies exerting a substantial impact on the regular operation of high-speed trains were pinpointed. Corresponding optimization recommendations were then proposed for a specialized detection and maintenance scheme, emphasizing fault prevention and control. Finally, a random forest model was used to calculate the impact of distinct events on the number of delayed trains, considering the impact of emergencies on the delay of high-speed trains, the event type, duration, location, and time point of occurrence. Additionally, a predictive model for the severity of emergencies was established to forecast the trains that may experience delays due to high-speed train emergencies.

References

1. Jafarian, E., & Rezvani, M. A. (2012). Application of fuzzy fault tree analysis for evaluation of railway safety risks: An evaluation of root causes for passenger train derailment. *Proceedings of the Institution of Mechanical Engineers Part F Journal of Rail & Rapid Transit, 226*(1), 14–25.
2. Bearfield, G., & Marsh, W. (2005). Generalising event trees using Bayesian networks with a case study of train derailment. *Lecture Notes in Computer Science, 3688*, 52–66.
3. Antonio, P., Fabrizio, R., & Raffaele, A. (2010). Bayesian analysis and prediction of failures in underground trains. *Quality & Reliability Engineering International, 19*(4), 327–336.
4. Liu, H. C., et al. (2018). Improving risk evaluation in FMEA with cloud model and hierarchical TOPSIS method. *IEEE Transactions on Fuzzy Systems, 99*, 1–1.
5. Huang, J., Li, Z., & Liu, H. C. (2017). New approach for failure mode and effect analysis using linguistic distribution assessments and TODIM method. *Reliability Engineering System Safety, 167*, 302–309.
6. Wang, Y., et al. (2014). Research on design evaluation of high-speed train auxiliary power supply system the AHP. In Transportation Electrification Asia-Pacific.
7. Akman, G. (2015) Evaluating suppliers to include green supplier development programs via fuzzy c-means and VIKOR methods. *Computers & Industrial Engineering, 86*(C), 69–82.
8. Gul, M., & Guneri, A. F. (2016). A fuzzy multi criteria risk assessment based on decision matrix technique: A case study for aluminum industry. *Journal of Loss Prevention in the Process Industries, 40*, 89–100.

9. Mohsen, O., & Fereshteh, N. (2017). An extended VIKOR method based on entropy measure for the failure modes risk assessment—A case study of the geothermal power plant (GPP). *Safety Science, 92*, 160–172.
10. Zhao, J., et al. (2017). An extended VIKOR method using intuitionistic fuzzy sets and combination weights for supplier selection. *Symmetry-Basel, 9*(9).
11. Yue, C. (2016). A geometric approach for ranking interval-valued intuitionistic fuzzy numbers with an application to group decision-making. *Computers & Industrial Engineering, 102*, 233–245.
12. Liu, F., & Yuan, X. H. (2007). Fuzzy number intuitionistic fuzzy set. *Fuzzy Systems & Mathematics, 21*(1), 88–91.
13. Danqing, X.U., Chen, X., & D.O. (2016). Mathematics, multi-attribute decision-making method based on hesitant intuitionistic fuzzy linguistic set. *Journal of Huaibei Normal University.*
14. Zhang, L., et al. (2018). A dynamic and quantitative risk assessment method with uncertainties for offshore managed pressure drilling phases. *Safety Science, 104*, 39–54.
15. Fakhravar, D., et al. (2017). Security vulnerability assessment of gas pipelines using discrete-time Bayesian network. *Process Safety and Environmental Protection, 111*, 714–725.
16. Meel, A., & Seider, W. D. (2006). Plant-specific dynamic failure assessment using Bayesian theory. *Chemical Engineering Science, 61*(21), 7036–7056.
17. Hamaidia, M., Kara, M., & Innal, F. (2018). Probability and frequency derivation using dynamic fault trees. *Process Safety Progress, 37*(4), 535–552.
18. Liang, C. Y., et al. (2009). A method of multi-attribute group decision making with incomplete hybrid assessment information. *Chinese Journal of Management Science, 17*(4), 126–132.
19. Tian, Z.-P., Wang, J.-Q., & Zhang, H.-Y. (2018). An integrated approach for failure mode and effects analysis based on fuzzy best-worst, relative entropy, and VIKOR methods. *Applied Soft Computing, 72*, 636–646.
20. Qin, Y., et al. (2015). Dynamic risk assessment of metro station with interval type-2 fuzzy set and TOPSIS method. *Journal of Intelligent & Fuzzy Systems, 29*(1), 93–106.
21. Markovic, N., et al. (2015). Analyzing passenger train arrival delays with support vector regression, *56c*(Jul), 251–262.
22. Oneto, L., et al. (2017). Dynamic delay predictions for large-scale railway networks: Deep and shallow extreme learning machines tuned via thresholdout, *47*(10), 2754–2767.
23. Corman, F., & Kecman, P. (2018). Stochastic prediction of train delays in real-time using Bayesian networks. *Transportation Research Part C: Emerging Technologies, 95*, 599–615.
24. Zeng, Y., Cheng, F., & Jin, J. (2019). Railway standard design, a neural network prediction model for the duration of delay in the dispatching section, 063(003), 148–153.
25. Huang, P., et al. (2018). Random forest model for predicting the recovery time of delayed trains on the Wuhan-Guangzhou high-speed railway, *40*(249(07)), 6–14.
26. Nair, R., et al. (2019). An ensemble prediction model for train delays, *104*(JUL), 196–209.
27. Liang, X., et al. (2017). A fault diagnosis method for train control onboard equipment based on Bayesian networks, *2017*(8): 96–103.
28. Zhao, Y., et al. (2014). Research on fault diagnosis method for high-speed railway signal system onboard equipment based on Bayesian networks, *2014*(11): 48–53.
29. Hung, C. F., & Hsu, W. L. (2017). Influence of long-wavelength track irregularities on the motion of a high-speed train. *Vehicle System Dynamics, 12*, 1–18.
30. Fu, Y., et al. (2017). Operation safety assessment of high-speed train with fuzzy group decision making method and empirical research. In *International Conference on Cloud Computing and Internet of Things.*
31. Yang, Y., Liu, Y., Zhou, M., Li, F., & Sun, C. (2015). Robustness assessment of urban rail transit based on complex network theory: A case study of the Beijing subway. *Safety Science, 79*, 149–162.
32. Li, Q., Song, L., List, G. F., Deng, Y., Zhou, Z., & Liu, P. (2017). A new approach to understand metro operation safety by exploring metro operation hazard network (MOHN). *Safety Science, 93*, 50–61.

33. Lin, S., Jia, L., Wang, Y., Qin, Y., & Li, M. (2016). Reliability study of Bogie system of high-speed train based on complex networks theory. In *International Conference on Electrical and Information Technologies for Rail Transportation 2016*.
34. Vilanova, M., et al. (2019). Assessment fertigation effects on chemical composition of Vitis vinifera L. cv. *Albarino. Food Chemistry, 278*, 636–643.
35. Song, Y., et al. (2018). Multivariate linear regression model for source apportionment and health risk assessment of heavy metals from different environmental media. *Ecotoxicology and Environmental Safety, 165*, 555–563.

Chapter 3
Service State Identification of Key Train Components on the Basis of Artificial Intelligence and Signal Processing

3.1 Overview

The mechanical parts of the train are bound to have distinct degrees of wear, fatigue, deformation, or other damage during their use. The technical condition of the mechanical components of the train will progressively degrade, leading to a reduction in service performance. The failure of these mechanical elements poses risks such as casualties, and property losses, and significantly impacts the operation of the rail transit system. Consequently, the condition assessment of components is the basis for ensuring the safe operation of rail transit. In addition, the realization of a component predictive maintenance strategy cannot be separated from accurate and efficient component condition assessment. The construction of a train intelligent perception network provides extremely convenient conditions for component condition monitoring.

Currently, vibration acceleration sensors are typically employed at home and abroad to monitor the service status of mechanical components of rail transit trains including bearings, and gearbox train frames, and a series of research results including signal processing and machine learning have been achieved [1]. Signal processing has high requirements for the inspection and maintenance personnel, and requires professional knowledge, which is challenging to popularize in the actual operation process. Moreover, rail transit trains operate under complex conditions, and vibration signals are characterized by non-steady state, randomness and low signal-to-noise ratio, which poses significant challenges to signal processing. Furthermore, signal processing methods encounter limitations in effectively utilizing multi-sensor information at the data level. With the swift progress in computer technology and sensor technology, coupled with the acquisition and accumulation of extensive monitoring data, machine learning methods have gained increasing popularity. State identification on the basis of machine learning generally includes three steps which are feature extraction, feature selection and state classification [2]. To extract effective features, engineers have to spend a great deal of energy on the condition that utilizing

Y. Qin et al., *Intelligent Safety and Reliability Analysis of Rail Transit Trains*,
https://doi.org/10.1007/978-981-97-4170-0_3

traditional machine learning methods, and the accuracy of state recognition depends largely on the merits of features.

State Identification of Training Key Components on the Basis of Artificial Intelligence.

Generally speaking, the state identification method of training key components on the basis of artificial intelligence should initially extract fault features from the collected original signals. Subsequently, effective feature selection can be conducted as required. Ultimately, pattern recognition shall be conducted for each type of fault signal to confirm its fault type. As the core of intelligent diagnosis technology, pattern recognition serves as an essential tool to achieve accurate fault diagnosis. The widely employed fault pattern recognition algorithm includes supervised learning methods, including Support Vector Machine (SVM) and Neural Networks (NN), and unsupervised learning methods, including K-Nearest Neighbor (KNN) [3].

Serving as a traditional supervised machine learning technique, Support Vector Machine (SVM) addresses a linear classifier by optimizing the separation distance of each sample within the feature space created through kernel mapping, which solves the problems of high dimensions and small samples. Sun [4] proposes a fault diagnosis method that combines dual-tree complex wavelet packet feature extraction and SVM classification. In the first place, employing the dual-tree complex wavelet packet decomposition to reconstruct the signal. Secondly, utilizing the energy feature of the reconstructed signal as the input of the SVM and finally applying the artificial fish swarm algorithm to optimize the parameters of the SVM. Moreover, Liang [5] proposes a parameter optimization algorithm on the basis of the cuckoo search of Levi flight to enhance the accuracy of fault diagnosis, aiming at the problem that traditional SVM is prone to fall into local optimization during classification.

As the representative of an unsupervised learning algorithm, the KNN method uses the training data set to divide the feature vector space in accordance with the idea of distance measurement to realize pattern recognition. Furthermore, KNN has the advantages of high accuracy, insensitive to outliers and being independent of data labels. Chen [6] applies the weighted KNN classifier and mixed domain features to make the early fault diagnosis of rolling bearing. The experimental outcomes demonstrate that the weighted KNN algorithm successfully extracts low-dimensional sensitive features. On the basis of the asymmetric neighborhood function, Pandya [7] proposed an enhanced KNN algorithm that optimizes the feature selection during classification and shows a better fault diagnosis effect than the ordinary KNN.

Despite the fact that the above theoretical research on fault diagnosis has achieved certain results, the applicability and accuracy of each method are frequently unsatisfactory in the actual industrial situation. Besides, the pattern recognition method on the basis of machine learning theory needs a great deal of tag data. The essential elements of the train operate normally for extended periods, resulting in limited monitoring data acquired during fault states. This leads to highly repetitive information in the monitoring data and the absence of typical fault-related information. Consequently, this constraint hinders the widespread industrial application of intelligent fault diagnosis technology. However, the emergence of the transfer learning concept has introduced a novel solution to address this challenge.

Transfer learning is a new machine learning method that uses knowledge or methods to solve problems in similar research fields. Transfer learning relaxes the basic assumption of identical distribution in traditional machine learning, and aims to transfer existing knowledge to solve the learning problem of merely a small number of labeled or even unlabeled sample data in the destination field [8]. Currently, transfer learning has been increasingly implemented in text classification, image and video recognition, audio signal recognition, behavior pattern recognition, etc. Furthermore, Pan et al. [9] summarize and review the latest progress of transfer learning in classification, regression, and clustering, and focus on the relationship between transfer learning and domain adaptation. Pan et al. [10] propose Transfer Component Analysis (TCA). A dimensionality reduction framework for feature adaptation is established on the basis of domain adaptation, and it has demonstrated favorable outcomes in cross-domain indoor Wi-Fi localization and cross-domain text classification assignments. Conversely, Feng et al. [11] use the overall image data set of trees, and the emergence of deep learning algorithms effectively makes up for the defects of artificial design features in machine learning. The computational model, consisting of multiple processing layers, has the capability to learn abstract feature expressions from the raw data [12]. The deep learning algorithm has greatly enhanced the technical level in many fields, including speech recognition [13], visual object recognition [14], object detection [15, 16], text representation [17], drug development [18], genomics [19], etc.

Additionally, the deep learning algorithm is utilized in the health detection of mechanical components. Chen et al. [20] employ a Sparse Auto-Encoder (SAE) to extract representative features from the statistical values of bearing signals, and the health status is recognized by utilizing a deep confidence network. Besides, Mao et al. [21] propose a method of fault diagnosis using a spectrum and self-coding extremum learning machine. Additionally, the automatic de-noising encoder performs the de-noising learning on the unlabeled data in an unsupervised way, and takes advantage of the frequency spectrum of the vibration signal as the input to realize the fault diagnosis [22]. Moreover, literature [23] employs deep neural networks (DNN) to realize fault feature extraction and intelligent diagnosis, and the spectrum and automatic encoder preprocess DNN. Certain researchers [24, 25] employ feature fusion by inputting multi-domain statistical features, encompassing time-domain features, frequency-domain features, and time–frequency features, into SAE.

Nevertheless, the above research is not completely learning from the original data. In other words, the extraction of effective features through numerous automatic coders is based on the time–frequency characteristics of vibration signals and is not intended to eliminate the need for manually designed features. Feng [26] and Shao [27] respectively provide feature learning methods on the basis of distinct automatic encoders, without undergoing any signal preprocessing or artificial feature extraction, thus eliminating the dependence on signal processing technology and diagnostic experience. Nonetheless, due to the same number of input units and sample length, the network structure is huge and the calculation cost is high. Currently, the sampling frequency is progressively increased. Additionally, with the progress of detection technology, including the reduction of sensor prices and the emergence of multi-function sensors, intelligent diagnosis methods with multi-dimensional data processing capabilities have gained growing significance in contemporary times. Besides, Convolution Neural Network (CNN) is the widely utilized deep learning architecture for handling images and other multidimensional data [28]. The feature representation layers of CNN are able to be stacked to create a deep network, which can enhance its ability to model intricate structures within the data. Furthermore, it will provide more data adjustment space for the algorithm and help enhance the adaptability of the method and data. Deep CNN has made breakthroughs in the processing of images [29], video [30], and audio [31] with its powerful nonlinear performance and recognition ability. The ISO2372 standard document suggests that the measurement of the vibration speed of the bearing shell should be conducted in three orthogonal directions which are vertical, horizontal and radial. Consequently, the end-to-end model should possess a minimum capability for processing 3D data. Inspired by this, our goal is to establish an end-to-end model of mechanical equipment fault diagnosis on the basis of CNN. The model has the capability to extract features from the raw data and handle high-dimensional data, enabling the automatic identification of mechanical components based on multi-source information.

Fault diagnosis method of training key components on the basis of signal processing method.

Rotating mechanical components including bearings and gearboxes play important parts in the operation of trains. Monitoring the condition and analyzing the performance degradation of the rotating machinery in trains is of paramount significance to ensure the safe operation of trains. The data formed in the condition monitoring and performance degradation analysis technology of rotating machinery include acoustic signals and vibration signals. Analyzing the degradation of railway train bearings in the early stage is performed based on acoustic signals [32–35]. Moreover, Ren [36] extracts the time-domain and frequency-domain characteristics of the vibration signal as the input of the deep learning network, as well as monitors the reliability of the bearing through the deep learning training method. Furthermore, Soualhi [37] extracts the time-domain characteristics of vibration signals, and evaluates the degradation status of bearings on the basis of a hidden Markov model and intelligent ant colony clustering. Utilizing the wavelet packet decomposition method, the vibration signal characteristics of the bearing are extracted, and the degradation performance of the bearing is evaluated employing self-organizing mapping and the Gaussian mixture model respectively [38, 39]. In accordance with the confidence value index, Hong [40] adopts the adaptive modal decomposition-self-organization mapping method to divide the degradation state of rolling bearings. Lei [41] applies a multi-dimensional hybrid intelligent algorithm to identify the degradation degree and category of the gearbox. Qu [42] employs a combination of Least Square Support Vector Regression (LSSVR) and genetic algorithm optimization to analyze mechanical vibration signals with noise, aiming to achieve risk assessment and machinery prediction. On the basis of the method of bearing fault diagnosis of high-speed train on the basis of the Moors spectral kurtosis diagram, Peng et al. [43] consider that the vibration signal from high-speed train bearings is susceptible to the impact of background noise and vibrations from other components under complex working conditions. Huang et al. [44] employ the method of blind source separation to separate the interference of bearing vibration signals, utilize blind source separation, the kurtosis of bearing vibration signals is taken as the sensitive feature for evaluating bearing performance, and establishing the early performance degradation model of bearing application dynamic fuzzy neural network.

The explosion of online data of mechanical equipment and the enhancement of the performance of hardware equipment have made deep learning applied and developed in the field of fault diagnosis. Moreover, Tamilselvan [45] applies the deep learning method to the fault diagnosis of aeroengine. Additionally, Althobiani et al. [46] utilize the deep learning method to identify the fault of the cylinder valve. Except for that, Tang et al. [47, 48] employ a deep learning algorithm to identify the fault type of high-speed train bogie. In the diagnosis of faults in rotating machinery, Shao et al. [49] employ an optimized deep-learning network model to identify bearing faults. Meanwhile, Zhang et al. [50] identify the fault of bearing under noisy environment by using CNN.

3.2 State Identification of Train Key Components on the Basis of Artificial Intelligence

3.2.1 Key Component Identification Method on the Basis of Shallow Transfer Learning

Traditional algorithms for component state recognition heavily rely on the learning capabilities of machine learning classification models. However, the optimal performance of these traditional machine learning methods often necessitates two preconditions: a substantial number of training samples and identical distributions between training and test samples. Nonetheless, it is challenging to obtain the actual fault samples of some key components in the industrial environment, which would result in a small sample problem. Simultaneously, due to the distinctions of the structure sizes, operating conditions and use environments of these key components, the data characteristics of their vibration signals will be distinct, which would lead to the same distribution conditions. Transfer learning, as a machine learning method designed for extracting shared knowledge between diverse domains, has the capability to alleviate distribution differences among samples from different domains. Consequently, it effectively addresses the aforementioned two issues, thereby improving the efficiency of cross-domain learning.

A comparison is made between traditional machine learning and transfer learning in Fig. 3.1. Figure 3.1 illustrates that traditional machine learning is limited to learning and recognizing within a single domain. Transfer learning, on the other hand, leverages knowledge from auxiliary domain learning to improve the recognition performance in the task domain. In other words, it involves transferring labeled samples or

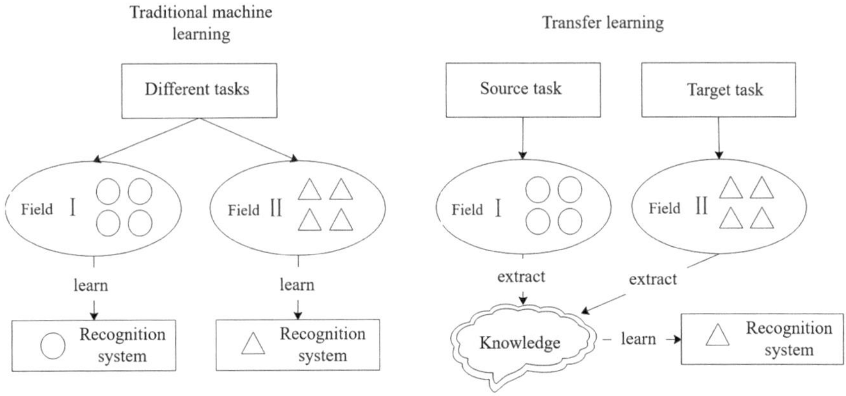

Fig. 3.1 Comparison of learning process between machine learning and transfer learning

feature structures associated with the task domain from the relevant auxiliary domain to bolster the recognition capability of the task domain.

3.2.1.1 Joint Distribution Adaptation and Enhancement of Basic Classifier KNN

As an essential research direction in transfer learning, domain adaptation is predominantly committed to solving the problem of distribution differences in distinct domains. The fundamental concept of domain adaptation involves mitigating the probability distribution disparities among domains through adaptation learning. The central challenge in domain adaptation is to minimize the distribution gap between the source and target domains. The selection of an appropriate metric for distribution difference holds significant importance. Moreover, the most commonly used measure of distribution difference in domain adaptation algorithms is the Maximum Mean Discrepancy (MMD). The principle of MMD is: Assumes that there is a mapping function $\phi()$ that can map the source field and destination field data to the same Reproducing Kernel Hilbert Space (RKHS), and the distribution difference is measured by the mean difference between the source field and the destination field in this space. Assumes that the source field sample is $\{x_1^s, x_2^s, \ldots, x_{n_s}^s\}$, the destination field sample is $\{x_1^t, x_2^t, \ldots, x_n^t\}$, MMD is able to be expressed as Eq. (3.1).

$$MMD_H^2(x^5, x^t) = \left\| \frac{1}{n_s} \sum_{i=1}^{n_s} \phi(x_i^5) - \frac{1}{n_t} \sum_{i=1}^{n_s} \phi(x_i^t) \right\|_H^2 \tag{3.1}$$

where $\| \|_H$ indicates the norm of the RKHS. After the square expansion of MMD_H, the inner product in RKHS is able to be equivalently converted into a kernel function and calculated by this kernel function.

Joint Distribution Adaptation (JDA) posits that effective domain adaptation should consider both adapting the marginal distribution and adjusting the conditional distribution. That is, $P(x_s) \neq P(x_t)$ and $P(y_s|x_s) \neq P(y_t|x_t)$. The target of JDA is to minimize the distance between the joint probability distribution of the source field and the destination field.

$$\text{distance}(D_s, D_t) = \|P(x_s) - P(x_t)\| + \|P(y_s|x_s) - P(y_t|x_t)\| \tag{3.2}$$

Due to the fact that the domain adaptation problem is mostly unsupervised learning, the label of the destination field is uncertain, that is, $P(y_t, |x_t)$ cannot be obtained. $P(y_t, |x_t)$ is typically replaced by class conditional probability $P(y_t, |x_t)$. The false label in the destination field y_t is acquired through utilizing KNN or SVM algorithm to train a classification model in the source field. It is determined that the Euclidean distance measure of traditional KNN classifier is not effective. Consequently, to fortify the KNN classifier, a more resilient correlation entropy distance measure is employed, leading to the establishment of the C-KNN classifier.

$X = (x_1, x_2, \ldots, x_n)^T$ and $Y = (y_1, y_2, \ldots, y_n)^T$ are two n-dimensional discrete vectors. The Correntropy Induced Metric (CIM) in the sample space is defined as Eq. (3.3).

$$CIM\,(X, Y) = (k_\sigma(0) - C_\sigma(X, Y))^{\frac{1}{2}} \tag{3.3}$$

where $C_\sigma(X, Y)$ indicates the correntropy between discrete variable X and Y. $k_\sigma(0)$ denotes the maximum value of correntropy under kernel width σ.

3.2.1.2 Algorithm Flow

For the purpose of making full use of the tagged data obtained in the laboratory environment and realizing the fault diagnosis of key components under the condition of small samples a key component migration diagnosis model under the condition of small samples, is constructed by integrating feature extraction (the extracted features are illustrated in Table 3.1, domain adaptation and C-KNN classification, and the process is demonstrated in Fig. 3.2.

3.2.1.3 Experimental Verification

Taking rolling bearing as an example, the effectiveness of the proposed method is verified in this section. This segment utilizes an ample supply of labeled data on rolling bearings from the Bearing Data Center of Case Western Reserve University. It combines this with the rolling bearing migration diagnosis model proposed in Sect. 3.2.4.2, which operates effectively under conditions of limited samples. This integration enables the diagnosis of faults in unlabeled freight car rolling bearings.

Table 3.1 Vibration signal characteristic name list

No	Characteristic	No	Characteristic	No	Characteristic
1	Peak value	9	Waveform factor	17	Root mean square frequency
2	Mean amplitude	10	Impulse factor	18	Skew frequency
3	Root-mean-square value	11	Clearance factor	19	Kurtosis frequency
4	Square root amplitude	12	Skewness factor	20	Shannon entropy
5	Absolute mean	13	Kurtosis Factor	21	Renyi entropy
6	Variance	14	Mean amplitude frequency	22	Power spectral entropy
7	Standard deviation	15	Gravity frequency	23	Approximate entropy
8	Peak factor	16	Standard deviation frequency	24	Energy index

(1) Algorithm Validity Test

The migration fault diagnosis data set selected in this section consists of two parts: ① The source field data set is composed of the public data set of Case Western Reserve University; ② Destination field data set is composed of freight car rolling bearing data.

The set of bearing data released by Case Western Reserve University comprises information pertaining to three different fault sizes, specifically measuring 0.007 inches, 0.014 inches, and 0.021 inches, respectively. Each of these fault sizes encompasses four unique load conditions, namely 0 horsepower, 1 horsepower, 2 horsepower, and 3 horsepower. Each load includes four health states which are normal state, inner ring fault, rolling element fault and outer ring fault. This section categorizes the data associated with all loads and health states within each fault size into domains denoted as S07, S14, and S21. Moreover, the sampling frequency is 12000 Hz. The sample length is set as 2048 consecutive sampling points, and the number of samples in each health state is 60. Consequently, the number of samples in the three source field datasets is 960.

The destination field dataset comprises data from SKF197726 freight car rolling bearings, representing four distinct health conditions denoted as T. The specific fault sizes include inner ring raceway stripping (10 mm × 45 mm), outer ring raceway stripping (10 mm × 30 mm), rolling body surface stripping (8 mm × 40 mm), and holder breakage. The collection conditions involve a rotation speed of 1000r/min, a vertical load of 272kN, a sampling frequency of 12800 Hz, a sample length of 2048 consecutive sampling points, and 200 samples under each health state. Consequently, the total number of samples in the destination field dataset is 800.

According to the aforementioned introduction, the migration diagnosis task in this section is set as S07 → T, S14 → T, and S21 → T. To fit the characteristics of missing marks in the freight car bearing data sample, the destination field data set T does not contain label information.

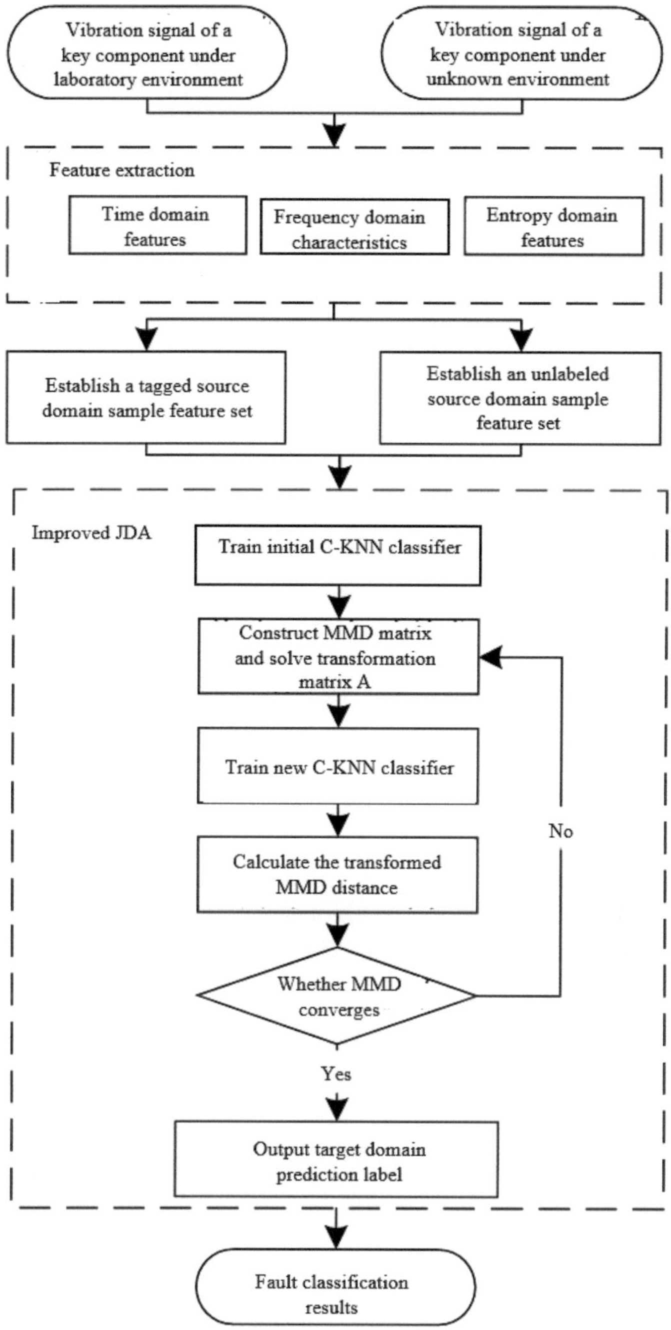

Fig. 3.2 Block diagram of key component migration diagnosis model under small sample condition

To verify the effectiveness of the proposed method, it is a comparison with classical classification algorithms KNN, SVM, enhanced KNN (C-KNN), and classical transfer learning algorithms including TCA and JDA. Based on the six classification methods described above, for three distinct migration tasks, the fault classification accuracy of each method and the average accuracy of three migration tasks are illustrated in Table 3.2.

Several conclusions can be drawn from Table 3.2. ① The transfer learning method is superior to the traditional classification method in distinct transfer tasks. The rationale behind this is that conventional classification methods lack domain adaptation, and the substantial distribution disparity between the source field and destination field leads to suboptimal classification outcomes. Consequently, transfer learning is an effective solution for fault diagnosis of rolling bearings with distinct operating conditions and models. ② Within the spectrum of traditional classification methods, C-KNN demonstrates superior overall performance, boasting the highest average classification accuracy. The possible reason is that some homonymous features of the source field and the destination field are quite distinct, and the local similarity measurement on the basis of the correntropy-induced distance has effectively served as a means of inhibition. ③ Among the three transfer learning methods, the diagnostic accuracy of the JDA and the proposed method is significantly higher than that of the TCA algorithm by over 10%. The fundamental reason is that the TCA only considers edge distribution adaptation, whereas the other two methods take both edge distribution and conditional distribution adaptation into consideration. Additionally, the method proposed in this study exhibits a marginal improvement over the JDA method. in S07 → T and T → S21 migration diagnosis tasks, and the average diagnostic accuracy is approximately 4% higher than the JDA method. On the one hand, in accordance with the C-KNN classification accuracy under the corresponding task, the reason is the advantage of the basic classifier. On the other hand, the proposed method changes JDA from edge distribution adaptation to conditional distribution adaptation to direct joint distribution adaptation. The aforementioned findings suggest that the migration diagnosis model under consideration is capable of effectively adapting to specific domains, mitigating disparities in distribution, performing unsupervised classification on freight car rolling bearing data using an adequate amount of labeled experimental data, and accomplishing the objective of rolling bearing migration diagnosis in small sample sizes.

Table 3.2 Comparison of accuracy of various methods under distinct migration tasks

Algorithm	$S_{07} \to T$	$S_{14} \to T$	$S_{21} \to T$	Average accuracy (%)
KNN	52.25	50.00	65.63	55.96
SVM	60.88	46.63	62.50	56.67
C- KNN	72.38	49.63	72.00	64.67
TCA	73.50	57.38	77.50	69.46
JDA	80.75	70.63	91.25	80.88
Shallow transfer learning	89.25	69.13	94.87	84.42

To further verify the effectiveness of the proposed migration diagnosis method, the source field samples and destination field samples of distinct migration learning methods are visualized. Besides, the source field and destination field samples are simultaneously dimensioned and mapped to two-dimensional space by the t-distribution Stochastic Neighbor Embedding (t-SNE) algorithm, and are indicated by distinct colors and shapes. Taking the migration diagnosis task T → S21 as an example, the visualization results following dimensionality reduction are illustrated in Fig. 3.3. Figure 3.3a reveals the original source field samples and destination field samples. Figure 3.3b demonstrates the results after TCA distribution adaptation. Figure 3.3c indicates the results following JDA distribution adaptation and Fig. 3.3d illustrates the distribution adaptation results of the proposed method. It is obvious in Fig. 3.3a that the distribution of samples of the same type in the source field and the destination field without distribution adaptation is quite distinct, and the effect of direct classification is poor. Figure 3.3b demonstrates that the TCA algorithm reduces the overall difference in data distribution. Nonetheless, due to the ignorance of the conditional distribution adaptation, the gap between the source field and the destination field samples of the same type is still large. Figure 3.3c illustrates that the JDA algorithm diminishes the condition distribution and edge distribution of samples in both the source and destination fields, owing to its incorporation of joint distribution adaptation. Nevertheless, certain roller failure and outer ring failure samples are still incorrectly classified. Figure 3.3d demonstrates that the proposed method also considers JDA, in the meantime, the number of samples misclassified is less than JDA. It can be concluded that the JDA method on the basis of the enhanced KNN algorithm is able to simultaneously adapt the edge distribution and the condition distribution, and the proposed method attains the highest efficiency in fault classification.

To further validate the efficacy of the proposed method in this study, the MMD distance and corresponding classification accuracy of various transfer learning methods in the iterative process are also computed, using the transfer diagnosis task T → S21 as an exemplar. The change of distribution adaptation distance of distinct transfer learning methods is demonstrated in Fig. 3.4a, and the change of corresponding classification accuracy is demonstrated in Fig. 3.4b.

The following conclusions can be drawn from Fig. 3.4. ① TCA method does not take into account conditional distribution adaptation, and consequently the MMD distance after adaptation is the largest. ② JDA method can reduce the distance between edge distribution and condition distribution simultaneously. Consequently, the JDA method has acquired obvious performance enhancement compared with the TCA algorithm. ③ The JDA method on the basis of the enhanced KNN reduces the distance between the source field and the destination field substantially. The main reason is that the JDA and the robustness of KNN enhanced by the basic classifier are conducted in the first step of the iteration.

Particularly, the reason why the proposed method cannot converge to a constant value in the iterative process is that the enhanced KNN of the basic classifier needs to re-estimate the kernel width of the correlation entropy distance measure in accordance

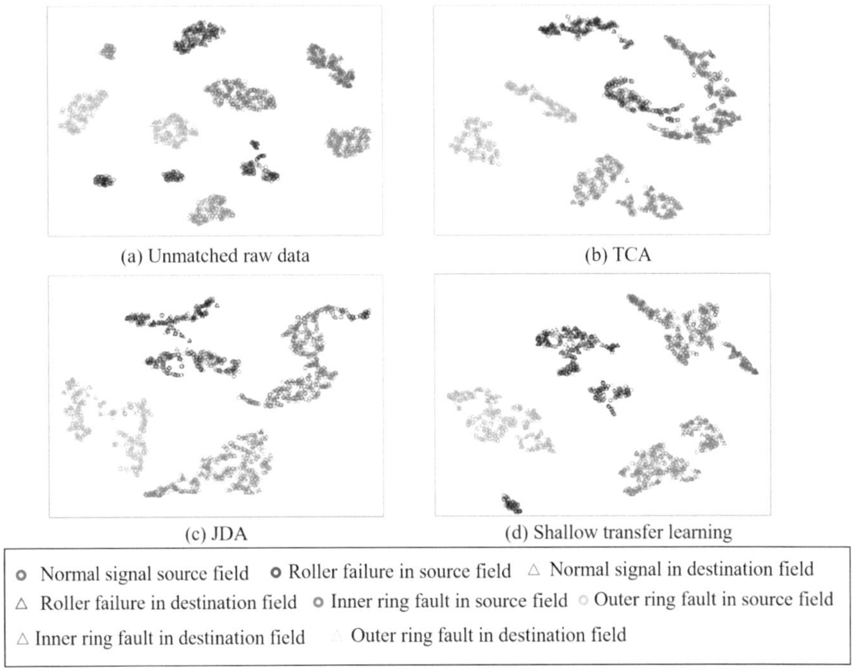

Fig. 3.3 Visualization of distribution adaptation results of distinct algorithms under migration task T → S21

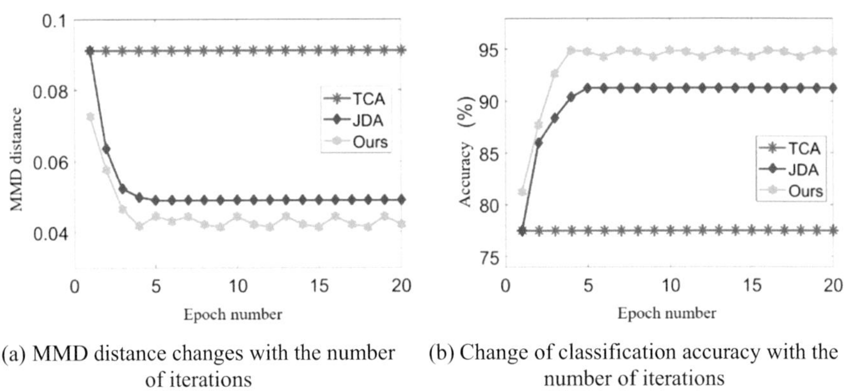

Fig. 3.4 MMD distance and corresponding accuracy change of domain distribution adaptation on task T → S21

with the training data during each classification, and thus there will be regular slight fluctuations in the final convergence process.

(2) Algorithm Parameter Sensitivity Test

Three diagnostic tasks are conducted to carry out sensitivity experiments of model parameters, and it is illustrated that the input parameters of the proposed method can maintain an excellent classification effect in a considerable range in this section.

Model regularization parameters λ: Fix other parameters unchanged, use distinct $\lambda \in [0.001, 100]$ value to analyze three migration diagnosis tasks, and the range of experimental parameters is set to λ. Theoretically, λ, it controls the complexity of the classification model. On the condition that $\lambda \to 0$, the penalty value of the classification model is extremely small, which contributes to the over-fitting problem of the model. As λ gradually increases, the degree of over-fitting decreases. When the value of λ approaches positive infinity, the underfitting problem arises from the classification model's failure to fully assimilate the internal laws of the data and the excessively large penalty term. Figure 3.5a illustrates the classification accuracy with parameters under distinct diagnostic tasks λ, from which we can determine that when $\lambda \in [0.001, 1]$, the classification effect of the model is comparatively stable and belongs to a reasonable value range.

Dimensionality reduction m: Fix other parameters unchanged analyze three migration diagnosis tasks with distinct m values, and set the experimental parameter range to $m \in [3, 24]$. In theory, m should choose a compromise value. A excessively small value for m may result in the adapted feature matrix being incapable of fully conveying the intrinsic information of the original data. Conversely, an excessively large value for m often entails a substantial computational load, leading to a loss of significance in dimensionality reduction.

Figure 3.5b indicates the change of classification accuracy with parameter m under distinct diagnostic tasks, from which we determine that $m \in [9, 24]$ is a reasonable

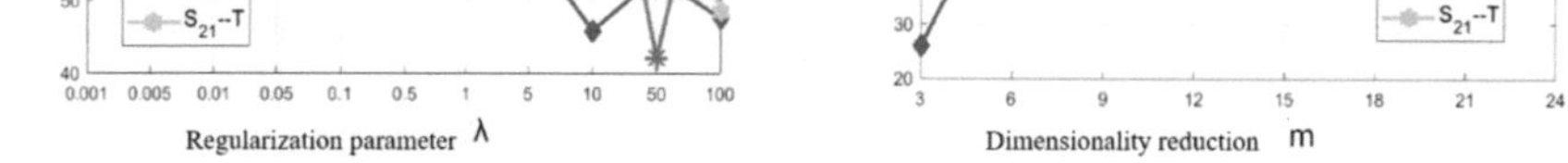

(a)The impact of regularization parameter λ on classification

(b)The impact of dimensionality reduction *m* on classification

Fig. 3.5 Classification performance changes of each migration task under distinct model parameters

value range. Within this range, the model's classification performance remains relatively stable, and the adaptively reduced feature matrix effectively encapsulates the information from the original feature matrix.

Consequently, m = 15 is generally taken in this experiment.

3.2.2 *State Identification Method of Key Components on the Basis of Multi-source Information and CNN*

3.2.2.1 Basic Concepts

(1) Tensor Expression and Information Fusion

A tensor is a natural extension of a vector and matrix. While William Rowan Hamilton introduced the concept of a tensor in 1846, its modern interpretation was later articulated by Waldmar Vogt in 1899. Einstein expounded general relativity with tensors in 1916, and then tensor analysis began to be valued and evolved into a potent mathematical tool. It finds extensive application in the representation of analytical mechanics, fluid mechanics, geometry, and electromagnetic theory and especially continuum mechanics. A tensor is a multilinear mapping defined on the Cartesian product of some vector spaces. The coordinate of a tensor is a quantity with |n| components in |n| dimensional space, where every component is a function of the coordinates.

Making p, q, m, n positive integers and $m, n \geq 2$. The (p, q)-order $(m \times n)$-dimensional rectangular tensor contains $m^p n^q$ components. $A = \left(a_{i_1,\ldots,i_p,j_1,\ldots,i_q}\right)$, $1 \leq i_1, \ldots, i_p \leq m$, $1 \leq j_1, \ldots, j_q \leq n$. When $p = q = 1$, the rectangular tensor A was a simple matrix $m \times n$.

Regarding data structure, tensors are multidimensional data, which can extend vectors and matrices to higher dimensions. Within the domain of artificial intelligence, tensors are utilized as multidimensional arrays, forming the foundation for numerous contemporary deep learning systems.

Information fusion is able to be divided into three levels: data layer information fusion, feature layer information fusion and decision layer information fusion.

Data layer information fusion involves the synthesis and analysis of the data directly obtained from the original data of various sensors without pretreatment. The primary advantage of data layer information fusion lies in its ability to preserve a maximal amount of field data, offering nuanced information that other fusion layers may not provide and potentially discovering more underlying data patterns.

The process of feature layer fusion is processing the original information, obtaining the extracted data features (statistical features or other empirical features), and then comprehensively analyzing and processing them. In comparison with data layer fusion, it attains a certain degree of information compression and is more suitable for real-time processing, and yet the effect relies on the effect of the proposed features.

Decision layer fusion is the fusion with the highest level, which directly provides the basis for control decision-making. Decision layer fusion has high flexibility, fast transmission, as well as outstanding fault tolerance. Nevertheless, decision layer fusion substantially depends on the accuracy of the original information preprocessing results. If the processing is incorrect, the associated costs would be substantial.

Tensors are widely employed in deep learning and have natural advantages in data fusion. A high-quality core data set is able to be obtained by representing big data with complex structure into a unified, concise, and efficient mathematical model, and adopting a certain dimensionality reduction algorithm, which is of great significance to the research of big data.

(2) CNN

A typical Convolutional Neural Network (CNN) typically comprises an input layer, multiple convolutional layers, activation layers, pooling layers, and fully connected layers.

(1) Input Layer

As illustrated in Fig. 3.6, preprocessing the original data includes zero-centered data (centralize all dimensions of the input data to 0), normalized data (normalize the data range to the same range to reduce the interference arising from the difference in the value range of each dimension data), as well as whitened (normalize the amplitude on each characteristic axis of the data).

(2) Convolutional Layer

As the core of CNN, the convolution layer is equivalent to the filter or feature extraction layer. Distinct features are extracted through local perception and weight sharing. The first convolution layer only extracts some low-level features, including edges, lines and corners. Moreover, subsequent convolution layers can iteratively acquire more complex features from the features extracted from the low-level convolution layer, Ultimately, the dimensionality reduction of the input high-dimensional data and the automatic extraction of the core features of the original data are realized.

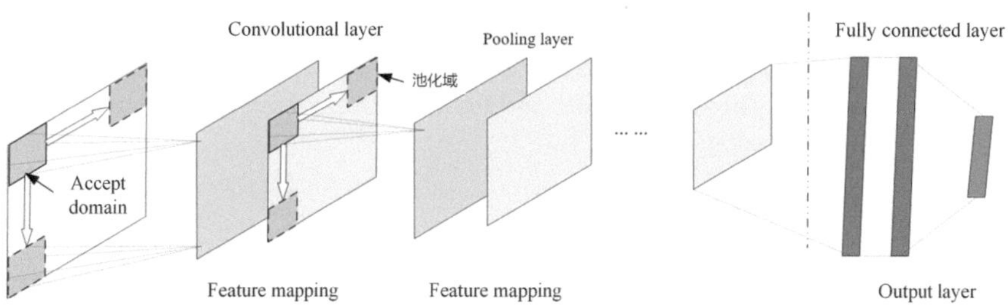

Fig. 3.6 Structure diagram of CNN

(3) Activation Layers

The activation layer employs a nonlinear activation function to process the linear output from the preceding layer, enhancing the network's capacity for nonlinear representation. In the field of deep learning, the Rectified-Linear Unit (ReLU) function is the activation function that is currently employed more frequently. ReLU function converges faster and does not lead to gradient disappearance.

(4) Pooling Layer

The pooling involves leveraging the local correlation of data within a network layer to achieve the function of downsampling. This process preserves essential information while reducing the overall size of the data.

Pooling layer lies in the middle of continuous convolution layers, which is applied to compress the amount of data and parameters and reduce over-fitting. Besides, max pooling which takes the maximum value in a certain rectangular area to reduce the sampling of data is the most prevalent pooling layer. Intuitively, the relative position of one feature to other features is more essential than its precise position. While diminishing the dimensions of the data space, the pooling layer will significantly decrease the number and computational load of network parameters. Rational sampling also has local linear transformation invariance, thus enhancing the generalization processing ability of convolutional neural networks and controlling over-fitting. In general, the pooling layer is typically interleaved between convolution layers to create a fundamental processing sequence known as "convolution-activation-pooling." Following the pure data feature extraction, the data dimension has also been reduced to the "fully connected" network processing, and the final output result is more controllable.

(5) Fully Connected Layer Additional Neural Network Neurons.

All neurons between the two layers have weight connections, which are typically set at the tail of the convolutional neural network. The fully connected layer is equal to the Multi-Layer Perceptron (MLP).

CNN has the outstanding fault-tolerant ability, parallel processing ability and self-learning ability. CNN can handle problems with complex environmental information, unclear background knowledge and unclear reasoning rules, as well as allows the sample to have large defects and distortion. CNN exhibits a relatively rapid operation speed. Furthermore, it possesses commendable adaptive capabilities, high resolution, and notably superior generalization abilities compared to alternative methods.

3.2.2.2 End-To-End Component State Identification Method on the Basis of Multi-source Information and CNN

Traditional component state recognition algorithms depend more on the individual experience and design features of engineers. Consequently, the qualities of these algorithms are difficult to guarantee. Moreover, it is challenging to effectively integrate and utilize the multi-source information provided by multi-sensors. Furthermore, the multi-source information fusion method on the basis of CNN is able to effectively avoid the complexity of artificial design features. By associating and synthesizing data and information from various sources, this method can comprehensively and promptly assess component status information. It achieves end-to-end component status identification and furnishes essential technical capabilities for early warning of component performance and reliability evaluation.

The component state identification method is on the basis of multi-source information fusion and CNN first conducts preliminary normalization processing on multi-source signals to reduce noise interference and speed up program convergence. Tensor expression is employed to fuse multi-source signals at the data layer subsequently. Finally, CNN is applied to realize the automatic identification of components.

(1) Data Layer Fusion and Loss Function Enhancement

With the rapid advancement of sensing technology, an increasing number of equipment monitoring samples encompass multiple sensor data or data from sensors in various directions. For instance, when monitoring the bearing, the vibration acceleration signal in two or three directions is typically used, and sometimes the temperature data is also monitored simultaneously. Consequently, fully integrating and utilizing all monitoring data at the data layer increases the likelihood of discovering new regular features. Thus, tensor expression is selected for data layer fusion.

Before that, it needs to be normalized to eliminate the negative impact of distinct data levels due to the distinct dimensions existing in the data.

$$x_i' = \frac{x_i - \min(x)}{\max(x) - \min(x)} \tag{3.4}$$

where $x = (x_1, x_2, \ldots, x_n)$ denotes monitoring data sample. x_i' indicates the i-th monitoring data of sample x. x_i' represents the normalization result of the i-th monitoring data.

There are N groups of monitoring data $s_1(t), s_2(t), \ldots, s_N(t)$, where $t = (t_1, t_2, \ldots, t_n)$. $s_1(t_i), s_2(t_i), \ldots, s_N(t_i)$ is the value at the time point t_i. After normalization according to Eq. (3.4), it is recorded as $S_1(t_i), S_2(t_i), \ldots, S_N(t_i)$.

The tensor of data at time point t_i is expressed as $T(t_i) = [S_1(t_i), S_2(t_i), \ldots, S_N(t_i)]$. Besides, the tensor expression samples employed for subsequent algorithm training are $T(t) = [T(t_i)], t = (t_1, t_2, \ldots, t_n)$. To facilitate the convolution of the neural network, the tensor expression sample will

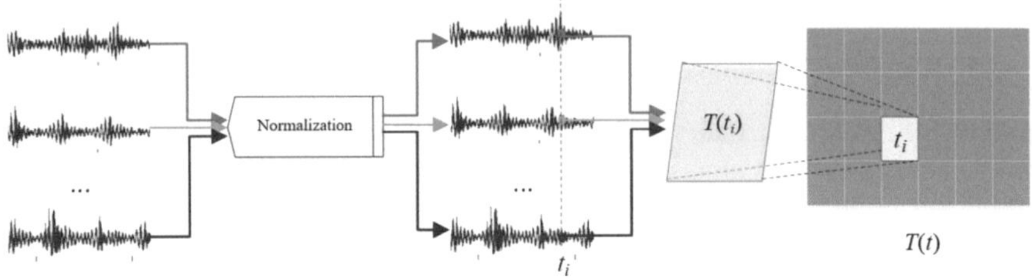

Fig. 3.7 Sample normalization and tensor expression

be converted from a single time dimension to a plane dimension, as demonstrated in Fig. 3.7.

It is determined that the convergence of the CNN is faster on the condition that the composite loss function is employed than the single loss function. Consequently, the mean value of the log-likelihood loss and the cross-entropy error loss is utilized as the loss function of the CNN.

(2) Algorithm Flow

The overall algorithm flow is demonstrated in Fig. 3.8. An 11-layer CNN is employed in this research.

The main model parameters are listed in Table 3.3, and the size of the model input depends on the number of sensor data as well as the size of the deformed plane.

3.2.2.3 Experimental Verification

Rolling bearing is an essential part of rotating machinery equipment [51]. Affected by a harsh working environment, it is extremely vulnerable to distinct kinds of damage, and results in halt, economic losses and additional severe repercussions [52]; Particularly in engines, approximately 41% of failures are arising from bearings [53]. Consequently, this section takes rolling bearing as an example to verify the effectiveness of the proposed method.

This experiment uses 6205-2RS JEM SKF bearing vibration acceleration data [54] provided by the Bearing Data Center of Case Western Reserve University, and its experimental device is demonstrated in Fig. 3.9. The main experimental equipment includes a 1.49 kw motor, torque encoder and power meter. The test data comprises vibration acceleration data for the four bearing states: Normal (N), Outer Ring Fault (ORF), Inner Ring Fault (IRF), and Rolling Element Fault (REF), with a sampling frequency of 12 kHz. The effectiveness of the algorithm and the robustness of the algorithm under distinct fault conditions are tested in the experiments.

Algorithm Validity Test

(1) Experimental Data Description and Preprocessing

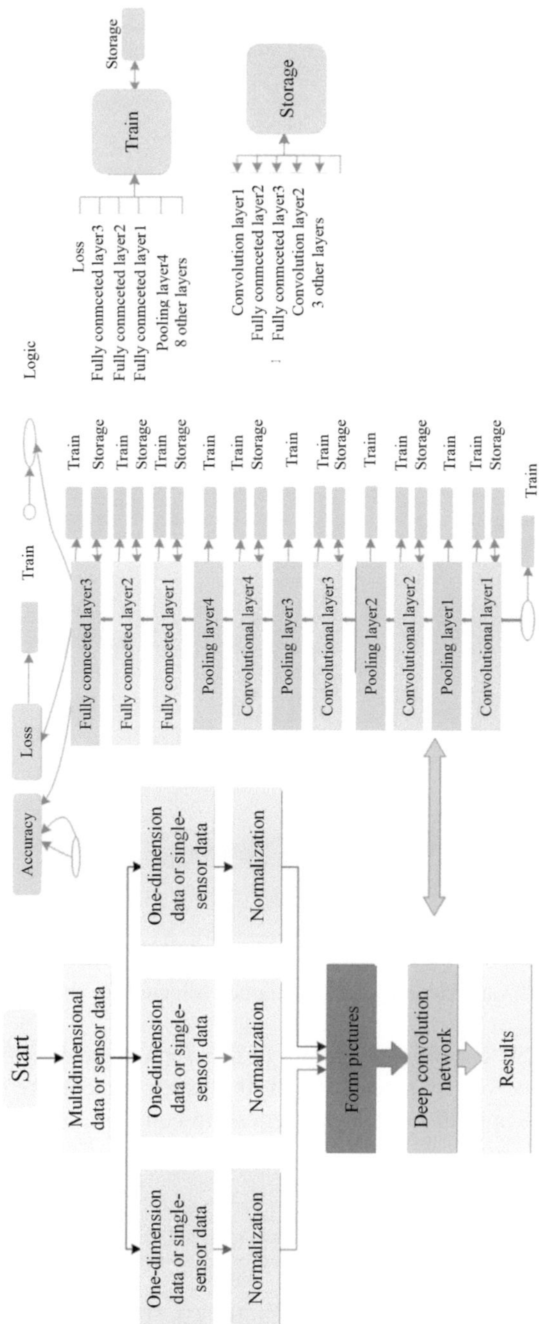

Fig. 3.8 Multi-dimensional end-to-end CNN model

Table 3.3 CNN model parameters

Layer	Size
Input	(None, 128, 128, 3)
Convolution layer 1	[5, 5, 3, 32]
Pooling layer 1	[1, 2, 2, 1]
Convolution layer 2	[5, 5, 32, 64]
Pooling layer 2	[1, 2, 2, 1]
Convolution layer 3	[3, 3, 64, 128]
Pooling layer 3	[1, 2, 2, 1]
Convolution layer 4	[3, 3, 128, 128]
Pooling layer 4	[1, 2, 2, 1]
Fully connected layer 1	[6 × 6 × 128, 1024]
Fully connected layer 2	[1024, 512]
Fully connected layer 3	[512, 4]

Fig. 3.9 Experimental device

The bearing data with fault degree of 0.007 inches selected for the test are listed in Table 3.4.

The files in the table contain three groups of data such as the drive end, the load end, and the reference vibration acceleration. In this experiment, the data set W_D_0 and W_D_1 are applied to perform the test. In comparison with single data,

Table 3.4 Data description

Dataset	Load (hp)	Speed (r/min)	IRF	REF	ORF	N
W_D_0	0	1797	IR007_0	B007_0	OR007@6_0	Normal_0
W_D_1	1	1772	IR007_1	B007_1	OR007@6_1	Normal_1
MIXED	W_D_0 + W_D_1					

multi-source information can provide more useful information, and substantially enhance the accuracy of the algorithm. To verify this conclusion, this subsection tests the acceleration data of the drive end, the drive end plus the load end, the drive end and the load end plus the reference vibration respectively, and names them One_test, Two_test and Three_Test experimental data. The data is segmented into 1024 points in each segment. Following normalization, it is represented as a three-dimensional tensor with dimensions 128 × 8 × 3. Each dimension encapsulates the information representation of the bearing status at different positions. The number of samples for each data set and distinct failure modes is demonstrated in Table 2.3.

To more intuitively present the differences between distinct sample data, it can be represented in the form of an image due to the fact that the experiment uses three-dimensional sample data. That is, the information on distinct dimensions corresponds to the RGB value of the image one by one, and on the condition that the dimension is missing, it is replaced by 0. Figure 3.10 demonstrates the image representation of IRF, REF, ORF, and N. Due to the absence of reference vibration acceleration data for normal state fault data, the two sub-graphs in the lower right corner of Fig. 3.5 are identical (N_Two_test and N_Three_test).

(2) Accuracy Comparison of Distinct Working Conditions with the Same Fault Degree

In the experiment, the accuracies of the one-dimensional to three-dimensional tensor samples from two sets of data sets W_D_0 and W_D_1 are tested. The experiment employs the 5-fold cross-validation method to calculate the average accuracy, with a maximum of 100 iterations and a learning rate of 0.001. The results are demonstrated in Fig. 3.11, where the red line, blue line, and green line represent 3D sample data, 2D sample data, and one-dimensional sample data respectively. Furthermore, the solid

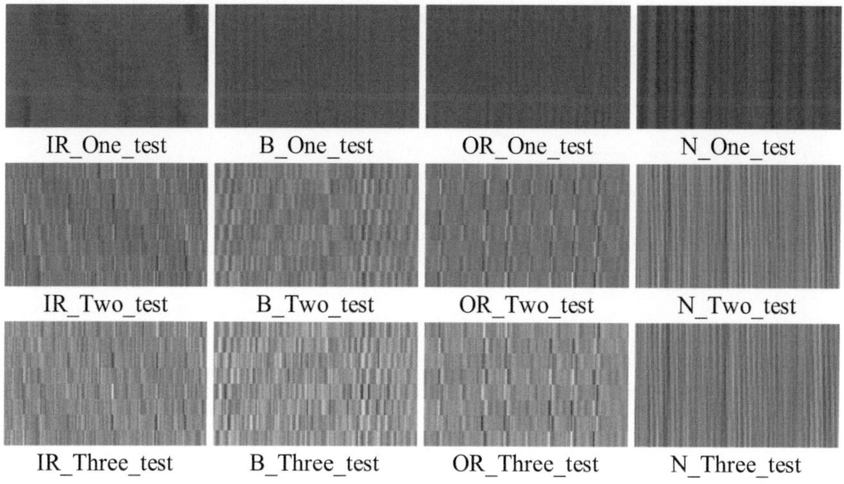

Fig. 3.10 Visual representation of dataset W_D_0

line and the dotted line represent the change in algorithm accuracy in the training stage and the test stage respectively.

Figure 3.11 indicates that the accuracy of this method is higher than 98% under distinct working conditions and sample sets, which effectively verifies the effectiveness of the method in this chapter. Additionally, Fig. 3.11a, b achieves 100% accuracy in the case of three data sources. During both training and testing, the three-source signal data achieved a stable state with the highest accuracy and the fewest iterations, effectively enabling the identification of bearing states. This underscores that the proposed multi-source end-to-end state identification algorithm, based on CNN, effectively leverages multi-sensor information, leading to highly accurate and efficient bearing state identification. Even in Fig. 3.11b, the two-dimensional tensor data sample has ten more iterations than the one-dimensional tensor sample, and yet the accuracy rate has increased from 98.33% to 99.16%. Given the swift advancement

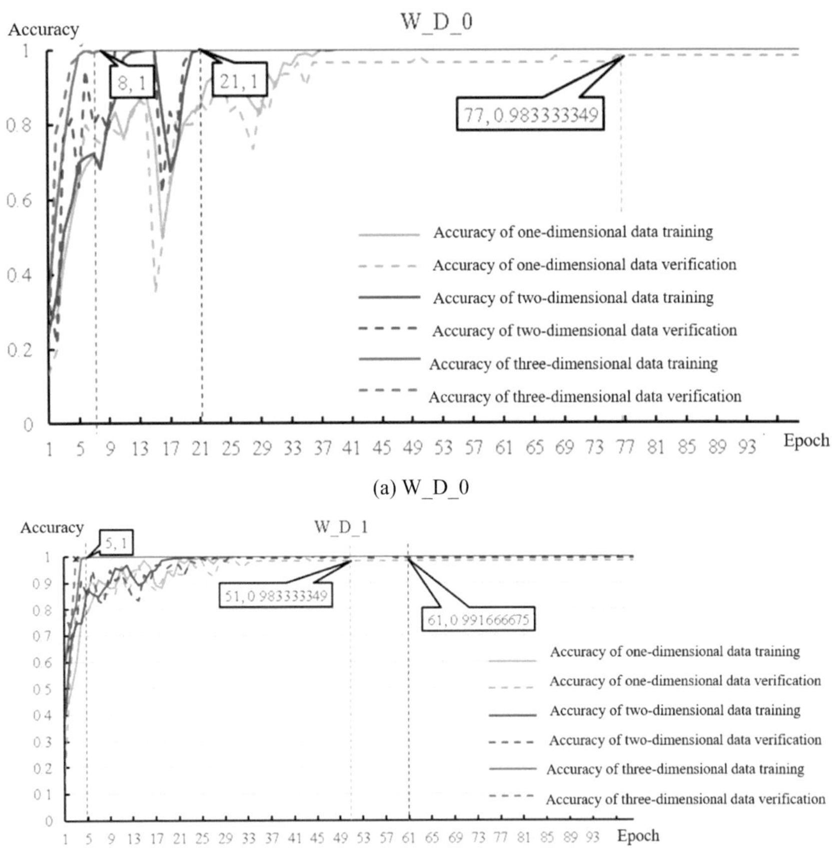

(a) W_D_0

(b) W_D_1

Fig. 3.11 Comparison of algorithm accuracy results

of computing power today, it becomes more meaningful to improve the accuracy rate by reducing the time required for an increased number of iterations.

(3) Accuracy Comparison of Mixed Working Conditions with the Same Fault Degree

In rail transit trains, particularly urban rail transit trains, the bearing rotation speed and load frequently change due to frequent acceleration and deceleration and the influence of passengers getting on and off. Moreover, traditional bearing state identification methods typically regard the bearing state under distinct working conditions (i.e., speed, load, etc.) as a new category during training and testing [55]. Nonetheless, the actual conditions change at any time and it is challenging to list them one by one. Consequently, it is urgent to put forward an effective method for bearing state identification under multiple conditions. In this section, datasets W_D_0 and W_D_1 are combined to create a mixed dataset, referred to as MIXED dataset, as demonstrated in Table 3.5 containing two speed and load cases to verify the accuracy of the proposed method.

As demonstrated in Fig. 3.12, under mixed conditions, the model proposed in this chapter still presents high accuracy in bearing fault state identification. Through the integrated utilization of three types of sensor information, the proposed model achieved a 100% accuracy rate and stabilized by the tenth iteration. In the case of a little less information (two sensor data), the number of iterations only increased by eight times without affecting the accuracy of the algorithm. Even when applying information from a single sensor, the classification accuracy of the model proposed in this chapter remains over 99%, which completely validates the outstanding performance of the multi-source end-to-end state identification model on the basis of CNN under the mixed conditions of the identical fault degree.

(4) Algorithm Efficiency Analysis

Table 3.5 Sample size of data set

Test		Number of IRF samples	Number of REF samples	Number of ORF samples	Number of N samples	Number of total samples
W_D_0	One_test	118	119	119	238	594
	Two_test					
	Three_test					
W_D_1	One_test	119	118	119	472	828
	Two_test					
	Three_test					
MIXED	One_test	237	237	238	710	1422
	Two_test					
	Three_test					
Label		0	1	2	3	–

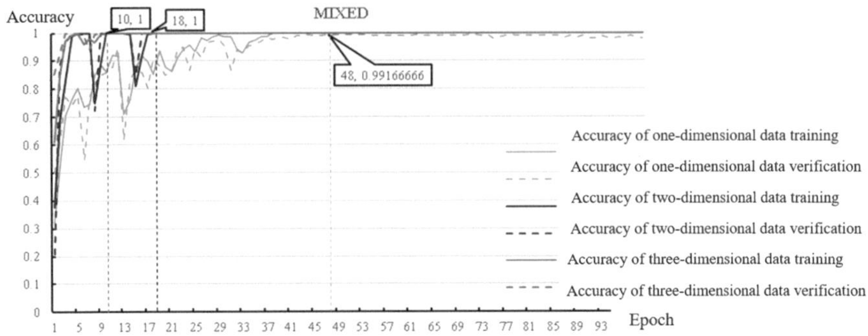

Fig. 3.12 Results of mixed data

This section has conducted experiments on the efficiency of multi-source end-to-end state identification algorithms. While keeping the signal processing, CNN structure, and other parameters unchanged, the algorithm's stopping condition is defined as reaching a loss value of 1.0E-04 or completing 100 iterations. The sample size of W_D_0, W_D_1 and MIXED are set as 594, 596, and 597, respectively.

The algorithm test environment is Intel (R) Core (TM) i7-4790 CPU @ 3.6 Hz, memory 8.00 GB, Samsung Solid State Drive 850 PRO 128 GB, WDC WD10EZEX-08M2NA0, NVIDIA GPU NVIDIA GeForce GTX 750.

Figure 3.13 demonstrates the relationship between the number of iterations and the target loss in the training and testing process of the proposed algorithm. An illustration of this can be seen in the convergence of the algorithm when presented with distinct information content and a mixed data set. As demonstrated by the red solid line and dotted line in Fig. 3.8, when the number of iterations of the algorithm exceeds five, the target loss of the proposed multi-source end-to-end state identification model on the basis of CNN tends to stabilize and reduce to the minimum. Nonetheless, the blue line and the green line gradually stabilized after nineteen and thirty-eight iterations respectively. Consequently, the algorithm proposed in this chapter indicates excellent performance in multi-source information processing. As information availability increases, so does the rate of convergence of the model.

Table 3.6 presents the minimum loss and time consumption results of the proposed algorithm under three data sets and three information volumes. The model introduced in this research attains the maximum accuracy with minimal loss and the shortest time when dealing with the largest volume of information. The algorithm's accuracy increases and the processing time decreases with a higher amount of data information. In the case of single-source and two-source sensor data, there is a need to strike a balance between time consumption and accuracy.

(5) Influence of Data Volume on Algorithm

Data volume is one of the key factors that affect the performance of modern nonlinear machine learning methods, particularly the performance of deep learning methods [56]. Figure 3.14 illustrates the relationship between data volume and algorithm

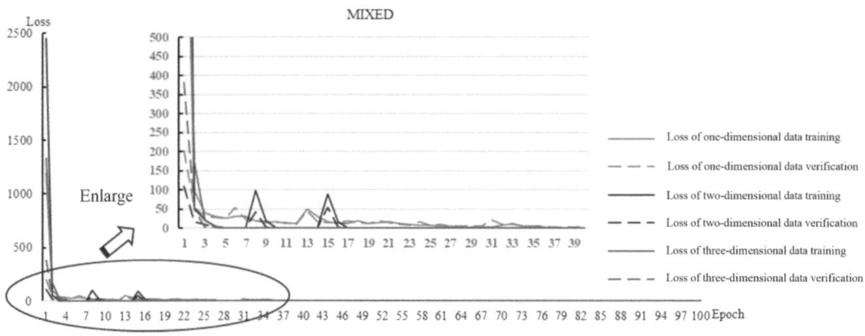

Fig. 3.13 Convergence comparison of algorithms under MIXED

Table 3.6 Comparison of algorithm efficiency

Data		Sample size	Number of training iterations	Loss		Accuracy		Time
				Train	Test	Train	Test	
W_D_0	One_test	594	100	0.0008	6.5871	1	0.9833	136.0765
	Two_test		100	0.0005	0.0007	1	1	199.2847
	Three_test		**56**	**1.05×10^{-5}**	**9.5×10^{-5}**	**1**	**1**	**75.0144**
W_D_1	One_test	596	100	0.000193	2.4752	1	0.9833	137.8747
	Two_test		100	0.000102	0.7449	1	0.9917	138.0615
	Three_test		**22**	**5.63×10^{-5}**	**8.9×10^{-5}**	**1**	**1**	**31.8715**
MIXED	One_test	597	100	0.000101	1.98543	1	0.9792	139.5175
	Two_test		91	1.67×10^{-5}	9.4×10^{-5}	1	1	126.2803
	Three_test		**41**	**6.43×10^{-5}**	**8.9×10^{-5}**	**1**	**1**	**57.9406**

performance proposed by Jason. Consequently, another experiment is conducted in this section to explore the impact of data volume on the performance of the multi-source end-to-end model on the basis of CNN.

In this experiment, data volumes of 100%, 50%, 25%, and 10% were selected from the MIXED dataset, denoted as F, H, Q, and 10%, respectively. These datasets were then tested under both three-dimensional and one-dimensional sample conditions. The four groups of data volumes are demonstrated in Table 3.7.

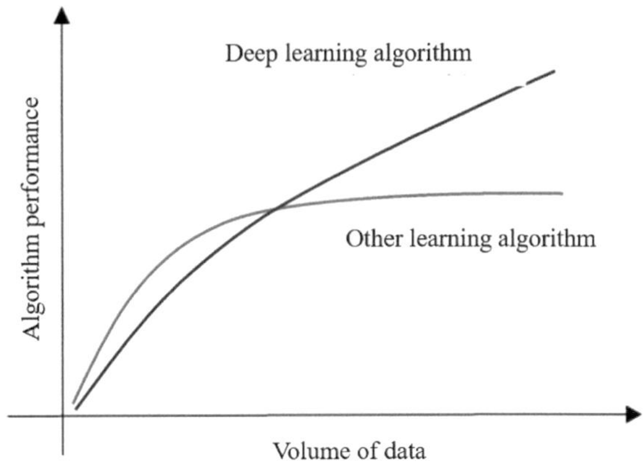

Fig. 3.14 Relationship between sample data size and algorithm performance

Table 3.7 Sample size of test data set

Proportion		100%	50%	25%	10%
MIXED	One-dimensional data	1422	711	355	142
	Three-dimensional data				

The results are presented in Fig. 3.15. Figure 3.15a indicates that the accuracy of the algorithm reaches the highest when the data volume is the largest, and decreases with the decrease of the data volume, which verifies Jason's conclusion. Figure 3.15b illustrates that with more information, a larger volume of data leads to faster algorithm convergence and reduced volatility during the convergence process. Consequently, in the case of a small amount of data, multiple methods can be employed to obtain multi-source information to enhance the accuracy of the algorithm. If it is challenging to obtain multi-source data, it is of great significance to increase the amount of data appropriately to enhance the performance of the algorithm.

The above experiments verify the effectiveness of the multi-source end-to-end state identification algorithm on the basis of CNN under the condition of the same fault degree, two similar and mixed conditions. Upon analyzing the information and data quantity, it is deduced that the proposed algorithm exhibits strong performance in the accuracy of bearing state identification. Moreover, the model achieves higher accuracy and faster convergence in scenarios with more effective information and abundant data.

(6) Algorithm Extension Test

The above experiments have conducted the bearing state identification of the algorithm under the identical fault degree, two similar and mixed conditions, and the experimental results are excellent. Nonetheless, the actual working conditions are more complex, and the range of speed and load changes is larger. Consequently, to further verify the robustness of this method in state identification, bearing fault data under distinct fault degrees and multiple mixed conditions are identified. The

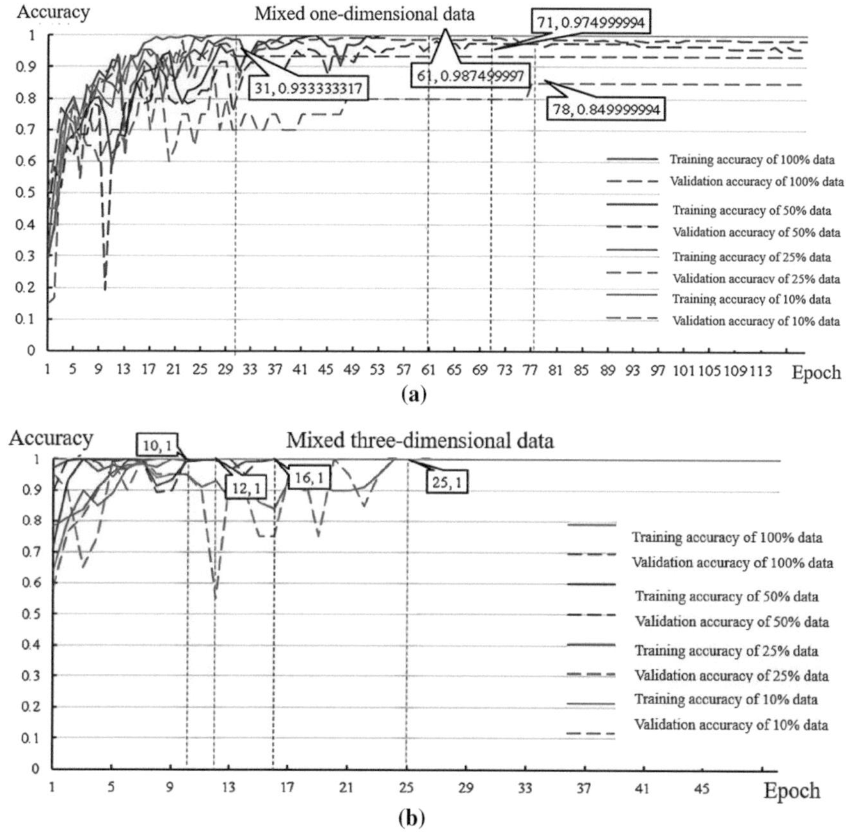

Fig. 3.15 Impact of data volume on model performance

experiment continues to utilize the dataset provided by the Bearing Data Center of Western Reserve University. The bearing failure degree is set as inches, 0.014 inches and 0.021 inches respectively. Moreover, the operating speed and load variation range are 1797 rpm to 1730 rpm, and 0 to 3 hp respectively, and the data used are demonstrated in Table 3.8.

Figure 3.16 indicates the 3D information data image representation of the file IR007_3, B007_3, OR007@6_3; IR014_2, B014_2, OR014@6_2; IR021_1, B021_1, and OR021@6_1. The image representation of 3D information data is completely distinct for data samples under identical fault modes and distinct working conditions, which is challenging for human eyes to effectively recognize.

The results are demonstrated in Fig. 3.17. The accuracy of the multi-source end-to-end model on the basis of CNN is 100% and 99.21% respectively in the training and testing process, and the algorithm convergence speed is fast. The experimental outcomes demonstrate that the proposed model maintains high accuracy and rapid

Table 3.8 Extended experimental data

Fault size	Load (HP)	Speed (r/min)	IRF	REF	ORF	N
0.007″	0	1797	IR007_0	B007_0	OR007@6_0	Normal_0
	1	1772	IR007_1	B007_1	OR007@6_1	Normal_1
	2	1750	IR007_2	B007_2	OR007@6_2	Normal_2
	3	1730	IR007_3	B007_3	OR007@6_3	Normal_3
0.014″	0	1797	IR014_0	B014_0	OR014@6_0	–
	1	1772	IR014_1	B014_1	OR014@6_1	–
	2	1750	IR014_2	B014_2	OR014@6_2	–
	3	1730	IR014_3	B014_3	OR014@6_3	–
0.021″	0	1797	IR021_0	B021_0	OR021@6_0	–
	1	1772	IR021_1	B021_1	OR021@6_1	–
	2	1750	IR021_2	B021_2	OR021@6_2	–
	3	1730	IR021_3	B021_3	OR021@6_3	–
Label			0	1	2	3

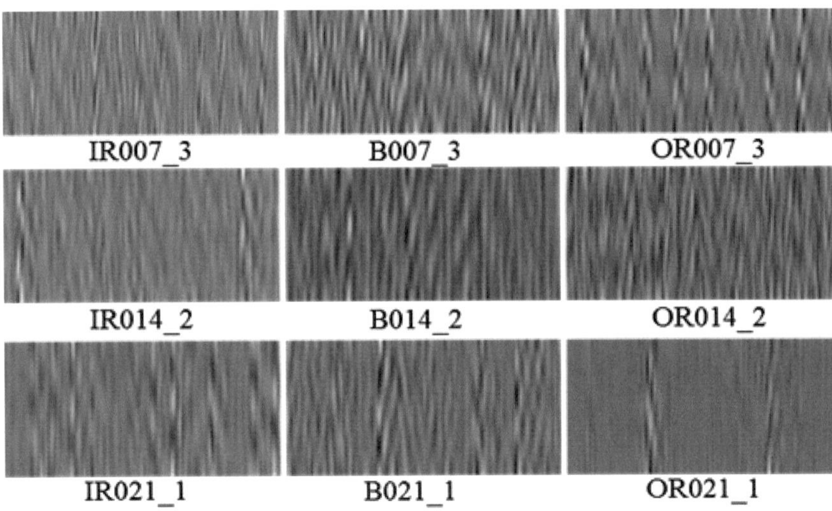

Fig. 3.16 Image representation of bearing fault samples under distinct working conditions

convergence speed even under various fault severities and complex operational conditions. This affirms its capability to offer dependable technical support for identifying bearing faults in real-world working conditions.

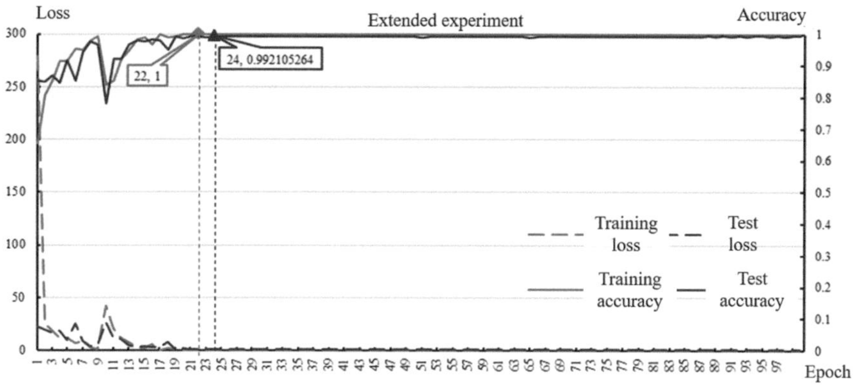

Fig. 3.17 Algorithm performance under complex conditions

3.3 Fault Diagnosis Method for the Key Components of Train on the Basis of Signal Processing Method

3.3.1 Basic Theory of Cyclostationary

3.3.1.1 Definition

The generalized cyclostationary process refers to a random process exhibiting specific periodicity in its statistical characteristics. If there exists a T in the probability density function of the random process that satisfies Eq. (3.5), the stochastic process is considered a strictly cyclostationary stochastic process.

$$p(x(t_1), x(t_2), ..., x(t_k)) = p(x(t_1 + L_1T), x(t_2 + L_2T), ..., x(t_k + L_kT)) \quad (3.5)$$

where $L_i(i = 1, 2, \ldots, k)$ is an arbitrary integer.

Typically, the researches of cyclostationary stochastic process are conducted in the sense of generalized cyclostationary, and consequently the generalized cyclostationarity is termed cyclostationarity for short.

3.3.1.2 Classification

The cyclostationary process can be categorized into first-order cyclostationary, second-order cyclostationary, and high-order cyclostationary based on the diverse statistical characteristics of the periodicity within the cyclostationary process.

In case the one moment $m_x(t)$ of the random process $x(t)$ satisfies Eq. (3.6), $x(t)$ is called first-order cyclostationary.

$$m_x(t) = m_x(t + T) \tag{3.6}$$

In case the autocorrelation function $R_t(x, \tau)$ of the random process $x(t)$ satisfies Eq. (3.7), $x(t)$ is termed second-order cyclostationary.

$$R_x(t, \tau) = R_x\left(t + T + \frac{\tau}{2}, u + T - \frac{\tau}{2}\right) \tag{3.7}$$

In case the k moment m_{kx} of the random process $x(t)$ satisfies Eq. (3.8), $x(t)$ is termed high-order cyclostationary.

3.3.2 *Cyclic Correntropy and Cyclic Correntropy Spectrum*

3.3.2.1 Definition of Cyclic Correntropy

Assuming that a correntropy function which satisfies Eq. (3.8) is periodic and its period is T_0.

$$V_x(t + T_0, \tau) = V_x(t, \tau) \tag{3.8}$$

Equation (3.8) is able to be converted into formula Eq. (3.9).

$$V_x(t, \tau) = \lim_{N \to \infty} \frac{1}{2N + 1} \sum_{n=-N}^{N} \kappa(x(t + nT_0) - x(t + \tau + nT_0)) \tag{3.9}$$

where N denotes data length, k represents kernel function.

In accordance with Fourier series, Eq. (3.10) can be obtained.

$$V_x(t, \tau) = \sum_{\alpha} V_x^{\alpha}(\tau) e^{j2\pi\alpha t} \tag{3.10}$$

where $\alpha = n/T$ $(n \in Z)$. α denotes cycle frequency or fault frequency related to fault signature of wheel set bearing. The coefficient of the Fourier series is illustrated in Eq. (3.11).

$$V_x^{\alpha} = \frac{1}{T_0} \int_{-T_0/2}^{T_0/2} V_x(t, \tau) e^{-j2\pi\alpha t} dt \tag{3.11}$$

The cyclic correntropy function of the signal $x(t)$ can be defined as the Fourier series of V_x^{α}, as demonstrated in Eq. (3.12).

$$V_x^{\alpha} = \lim_{T \to \infty} \frac{1}{T} \int_{-T/2}^{T/2} \kappa_{\sigma}(x(t), x(t + \tau)) e^{-j2\pi\alpha t} dt \tag{3.12}$$

where k_a represents kernel function.

3.3.2.2 Properties of Cyclic Correntropy

Cyclic correntropy spectral density function, which is abbreviated as cyclic correntropy spectrum, is the Fourier transform of cyclic correntropy, as demonstrated in Eq. (3.13).

$$S_x(\xi;f) = \int_{-\infty}^{\infty} V_x(\xi;\tau)e^{-j2\pi f\tau}d\tau \tag{3.13}$$

The projection function of the cyclic correntropy spectral density onto the cyclic frequency domain, abbreviated as cyclic correntropy spectral projection, is a function of the cycle frequency. When the cycle frequency is fixed, the value of the cyclic correntropy spectral projection is the maximum value of the cycle correntropy corresponding to the cycle frequency, and the definition formula is presented in Eq. (3.14).

$$P_x(\xi) = maxS_x(\xi;f), f \in R \tag{3.14}$$

where R denotes the set of real numbers.

3.3.3 Principles of Support Vector Machine

Support Vector Machine (SVM) proposed by Vapink is a machine learning method on the basis of the structural risk minimization principle. Support Vector Machines (SVM) demonstrate exceptional generalization and recognition capabilities, particularly for datasets with small volumes and high-dimensional nonlinear characteristics. The principle of SVM is to establish a hyperplane as a decision surface to maximize the isolation edge of distinct types of samples. SVM was initially utilized to address the linear binary classification problem, and the linear optimal hyperplane is established in accordance with the geometric interval of the support vector.

3.3.3.1 Optimal Classification Hyperplane Construction

The principle of the classification plane ought to be clarified before the in-depth description of the principle of SVM. Besides, the simplest classification plane is a linear classification plane. For a two-classification problem, the n-dimensional vector $\boldsymbol{x}$ is employed to symbolize a group of data samples, $\boldsymbol{\upsilon}$ is a parameter vector and $\boldsymbol{\upsilon}^{\mathrm{T}}$ is its transposition, $\boldsymbol{y}$ indicates the category attribute and each value range of its element is$[-1,1]$, representing two distinct categories. A hyperplane in an n-dimensional data space is defined by Eq. (3.15).

$$(\upsilon \cdot x) + b = 0 \tag{3.15}$$

The sample is linearly separable in case Eq. (3.15) satisfies Eq. (3.16).

$$\begin{cases} \langle \upsilon^T \cdot x \rangle + b \geq 1 & y = 1 \\ \langle \upsilon^T \cdot x \rangle + b \leq -1 & y = -1 \end{cases} \tag{3.16}$$

Let $f(x) = < \boldsymbol{\upsilon}^{\mathbf{T}} \bullet \boldsymbol{x} > + b$, if $f(x) = 0$, x locates on hyperplane H. If $f(x) \leq -1$ or $y = -1$, x locates on the H_1 side of hyperplane H. If $f(x) \geq 1$ or $y = 1$, x locates on the H_2 side of hyperplane H. The unified form illustrated in Eq. (3.17) can be acquired through Eq. (3.16).

$$y_i[(\upsilon \cdot x_i) + b] - 1 \geq 0 \quad i = 1, 2, \ldots, n \tag{3.17}$$

The samples in H_1 and H_2in Fig. 3.18 are the support training samples (i.e., Support Vectors, SV). To make the classification hyperplane distinguish the two types of samples to the maximum extent, it is necessary to maximize the margin between samples of the two classes. The optimal hyperplane is obtained by satisfying the optimal solution of the quadratic optimization problem as outlined in Eq. (3.18).

$$\begin{aligned} &\mathbf{min} \quad \Phi(\upsilon) = \frac{1}{2}\|\upsilon\|^2 \\ &s.t. \quad y_i[(\upsilon \cdot x_i) + b] - 1 \geq 0 \quad i = 1, 2, \ldots, n \end{aligned} \tag{3.18}$$

Nevertheless, the maximum-margin classifier consistently generates an assumption without training error. When the samples cannot be linearly separable, the

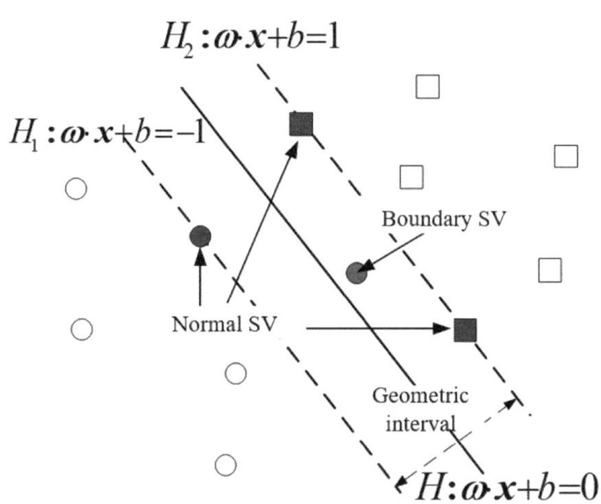

Fig. 3.18 Schematic diagram of optimal classification hyperplane

maximum geometric interval is less than 0, which is meaningless. Consequently, slack variables are typically introduced to allow individual samples to violate the interval constraint to a certain extent without influencing the overall classification outcome. Equation (3.18) transforms to Eq. (3.19).

$$\begin{cases} \mathbf{min}\Phi_1(\upsilon, \xi) = \frac{1}{2}\|\upsilon\|^2 + \gamma \sum_{i=1}^{n} \xi_i \\ s.t.\ \ y_i[\langle \upsilon \cdot x \rangle + b] \geq 1 - \xi_i \\ \xi_i \geq 0, \quad i = 1, 2, \ldots, n \end{cases} \tag{3.19}$$

where Φ_1 represents the objective function, ξ_i denotes the slack variables, and γ indicates the penalty coefficient. Lagrange function, which can obtain the optimal solution by solving the dual problem, is utilized to solve this convex optimization problems with linear constraints effectively. This is the SVM dual algorithm under linear separable condition, and the objective function expression demonstrated in Eq. (3.20) is obtained by introducing Lagrange multipliers to constraints α.

$$\underset{\alpha_i > 0}{max}\, L(\upsilon, b, \alpha) = \frac{1}{2}\|\upsilon\|^2 - \sum_{i=1}^{n} \alpha_i \big(y_i \big(\upsilon^T x_i + b\big) - 1\big) \tag{3.20}$$

The original problem is equivalent to Eq. (3.21) after dual transformation.

$$\begin{cases} \underset{\alpha,\beta}{\mathbf{max}}\, \underset{\upsilon,b,\xi}{\mathbf{min}} \left\{ L = \frac{1}{2}\|\upsilon\|^2 + \gamma \sum_{i=1}^{n} \xi_i - \sum_{i=1}^{n} \alpha_i \big[y_i(\langle \upsilon \cdot x_i \rangle + b) - 1 + \xi_i\big] - \sum_{i=1}^{n} \beta_i \xi_i \right\} \\ \text{s.t. } \alpha_i > 0 \beta_i > 0 \end{cases} \tag{3.21}$$

where α_i and β_i are Lagrange multipliers. The purpose of constructing Lagrange function is to acquire the minimum value of υ and b.

On the basis of the Karush–Kuhn–Tucker (KKT) conditions, the homogeneous partial derivatives of Eq. (13.19) to υ and b are calculated respectively in Eq. (3.22). The original function is transformed into the dual problem demonstrated in Eq. (3.23).

$$\begin{cases} \frac{\partial L}{\partial \omega} = \upsilon - \sum_{i=1}^{n} y_i \alpha_i \boldsymbol{x}_i = 0 \\ \frac{\partial L}{\partial \xi} = \gamma - \alpha_i - \beta_i = 0 \\ \frac{\partial L}{\partial b} = \sum_{i=1}^{n} y_i \alpha_i = 0 \end{cases} \tag{3.22}$$

$$\begin{cases} \underset{\alpha}{\mathbf{max}} \left[L_D = \sum_{i=1}^{n} \alpha_i - \frac{1}{2} \sum_{i,j=1}^{n} y_i y_j \alpha_i \alpha_j \langle x_i, x_j \rangle \right] \\ s.t.\ \ 0 \leq \alpha_i \leq \gamma, \\ \sum_{i=1}^{n} \alpha_i y_i = 0 \end{cases} \tag{3.23}$$

Finally, the results of the problem, which is demonstrated in Eq. (3.24), are obtained by utilizing Sequential Minimal Optimization (SMO) algorithm.

$$\begin{cases} \alpha_i = 0 & (1) \\ 0 < \alpha_i < \gamma & (2) \\ \alpha_i = \gamma & (3) \end{cases} \tag{3.24}$$

Generally, the outcomes for the majority of sample points correspond to (1). Only a relatively small number of results correspond to (2) and (3), and these sample points are Support Vectors (SV). The corresponding x_i in results (2) is the Normal SV (NSV), and the corresponding x_i in results (2) is the Boundary SV (BSV), that is, the wrong sample points. v and b is able to be obtained through calculating α_i, as demonstrated in Eq. (3.25).

$$\begin{cases} v = \sum_{i=1}^{n} \alpha_i y_i \cdot x_i \\ b = \frac{1}{N_{NSV}} \sum_{x_i \in NSV} y_i - \sum_{x_j \in SV} \alpha_j y_j \langle x_i \cdot x_j \rangle \end{cases} \tag{3.25}$$

The ultimate optimal classification function, as presented in Eq. (3.26), can be derived based on the aforementioned analysis.

$$f(x) = \mathbf{sgn}\left[\left\langle v^T \cdot x \right\rangle + b\right] = \mathbf{sgn}\left[\underbrace{\sum_{i=1}^{n} \alpha_i y_i \left\langle x_i^T \cdot x \right\rangle}_{\langle v^T \cdot x \rangle} + \underbrace{\frac{1}{N_{NSV}} \sum_{x_i \in NSV} y_i - \sum_{x_j \in SV} \alpha_j y_j \langle x_i \cdot x_j \rangle}_{b}\right] \tag{3.26}$$

3.3.3.2 Nonlinear Case

The majority of data are nonlinear and separable, and currently, there is no direct methodology to solve nonlinear problems. Generally, it is mapped into a high-dimensional linearly separable data space through mathematical transformation to calculate the optimal classification plane, and subsequently calculate the inner product of the optimal classification surface and training data points in accordance with Eq. (3.26).

Supposing a nonlinear sample set is $[(x_i, y_i)| \, i \in n]$. The set is transformed into another data space following mapping, in which it is linearly separable. The transformed set becomes $[(\Phi(x_i), y_i)| \, i \in n]$. Consequently, the corresponding classification function exists in the transformed data space as demonstrated in Eq. (3.27).

$$f(x) = \mathbf{sgn}\left(\sum_{i=1}^{n} \alpha_i y_i \langle \phi(x_i^T) \cdot \phi(x) \rangle + b\right) \tag{3.27}$$

The calculation of the inner product in the aforementioned equation does not need the specific form of $\phi(x)$, which saves the tedious calculation complexity and redundant data dimensions. Consequently, this represents an efficient solution for situations where the data is nonlinearly separable. The inner product of Eq. (3.27) is demonstrated in Eq. (3.28).

$$K(x_i \cdot x) = \langle \phi(x_i^T) \cdot \phi(x) \rangle \tag{3.28}$$

The optimal classification plane is presented in Eq. (3.29).

$$f(x) = sgn\left(\sum_{i=1}^{n} \alpha_i y_i K(x_i \cdot x) + b\right) \tag{3.29}$$

Parameter b is calculated through Eq. (3.30).

$$\frac{1}{N_{NSV}} \sum_{x_i \in NSV} y_i - \sum_{x_j \in SV} \alpha_{jy_j} K(x_i . x_j) \tag{3.30}$$

In accordance with statistical learning theory, it is not necessary to define the form of nonlinear transformation due to the fact that only inner product operation is required. Besides, the Hilbert–Schmit principle indicates that $K(x_i, x)$ must be a kernel function satisfying the Mercer condition. SVM solves the linear indivisibility problem in the original space by utilizing the kernel function. One of the advantages of SVM is that it avoids the increased computational complexity associated with nonlinear expansion. In the situation of non-separable linearity, SVM maps the input variables to a high-dimensional feature space via a specified kernel function, and constructs the optimal classification hyperplane in this high-dimensional space. To classify data sets using SVM, the input space is mapped to the high-dimensional feature space using the pre-selected kernel function initially. Currently, multiple kernel functions are widely used, and Table 3.9 lists the 14 most common kernel functions.

3.3.3.3 Least Squares Support Vector Machine

Least Squares Support Vector Machine (LSSVM) applies the structural risk principle to select the loss function which is distinct from SVM in the optimization objective. In other words, LSSVM employs the 2-norm of interval error constructed by the least square to replace the error 1-norm of SVM. Consequently, the quadratic convex optimization problem presented in Eq. (3.21) can be transformed into Eq. (3.31).

Table 3.9 Common kernel functions

Name	Function expression
Linear kernel	$K(x \cdot y) = x^t y$
Polynomial kernel	$K(x \cdot y) = \left(ax^t y + c\right)^d$
Gauss kernel	$K(x \cdot y) = \exp\left(-\frac{\|x-y\|}{2\sigma^2}\right)$
Exponential kernel	$K(x \cdot y) = \exp\left(-\frac{\|x-y\|}{\sigma}\right)$
Sigmoid kernel	$K(x \cdot y) = \tan\left(\alpha x^t y + c\right)$
Anova kernel	$K(x \cdot y) = \exp\left(-\sigma\left(x^k - y^k\right)^2\right)^d$
Rational quadratic kernel	$K(x \cdot y) = 1 - \frac{\|x-y\|^2}{\|x-y\|^2+c}$
Multiquadric kernel	$K(x \cdot y) = \left(\|x - y\|^2 + c^2\right)^{0.5}$
Wave kernel	$K(x \cdot y) = \frac{v}{\|x-y\|} \sin\left(\frac{\|x-y\|}{v}\right)$
Triangular kernel	$K(x \cdot y) = -\|x - y\|^d$
Log kernel	$K(x \cdot y) = -\log\left(1 + \|x - y\|^d\right)$
Spline kernel	$K(x \cdot y)=1+x^t y+x^t y \min(x, y)-\frac{x+y}{2} \min(x, y)^2 + \frac{1}{3} \min(x, y)^3$
Cauchy kernel	$K(x \cdot y) = \frac{1}{\|x-y\|^2/\sigma+1}$
Generalized t-student kernel	$K(x \cdot y) = \frac{1}{1+\|x-y\|^2}$

$$\begin{cases} \mathbf{min}\Phi_2(\upsilon, \gamma, e) = \frac{\mu}{2}\|\upsilon\|^2 + \frac{\zeta}{2}\sum_{i=1}^{n} e_i \\ s.t. \quad y_i[\phi(x_i) + b] = 1 - e_i \quad i = 1, 2, \ldots, n \end{cases} \tag{3.31}$$

where e_i denotes the quantized interval error. The function of e_i indicates to be introduced as a slack variable for the purpose that LSSVM can transform inequality constraints into equality constraints. Put differently, A QP problem is transformed into a problem of solving linear equations, through which substantially reduces the complexity of solving Lagrange multipliers $\alpha_i > 0$. The sum of square errors can be acquired by multiplying y_i by the slack variable e_i and on the basis of $y_i{}^2 = 1$, as demonstrated in Eq. (3.32).

$$\sum_{i=1}^{n} e_i^2 = \sum_{i=1}^{n} (y_i e_i)^2 = \left(y_i - \left(\upsilon^T \phi(x) + b\right)\right)^2 \tag{3.32}$$

where

$$e_i = y_i - \left(\upsilon^T \phi(x) + b\right) \tag{3.33}$$

Consequently, the objective function of LSSVM classifier can be simplified as Eq. (3.34).

$$\begin{cases} \Phi_2(\upsilon, b) = \mu E_\upsilon + \zeta E_D \\ E_\upsilon = \frac{1}{2}\upsilon^T\upsilon \\ E_D = \frac{1}{2}\sum_{i=1}^{n} e_i^2 = \frac{1}{2}\sum_{i=1}^{n}\left[y_i - \left(\upsilon^T\phi(x_i) + b\right)\right]^2 \end{cases} \tag{3.34}$$

where μ and ζ are hyperparameters and the function of these hyperparameters is to adjust the amount of regularization by the sum of relative square errors. Moreover, merely hyperparameter ratio γ is applied as tuning parameter in Eq. (3.34). Introducing μ and ζ will play an essential role in the Bayesian inference process of LSSVM loss function.

Likewise, Eq. (3.34) undergoes transformation into the Lagrange formula, as illustrated in Eq. (3.35).

$$\underset{\alpha}{max}\ \underset{\upsilon,b,e}{min}\left\{L = \Phi_2(\upsilon, b, \gamma, e) - \sum_{i=1}^{n}\alpha_i\left[y_i(\langle\theta\cdot x_i\rangle + b) - 1 + e_i\right]\right\}(\alpha_i \in R) \tag{3.35}$$

According to KKT conditions, the partial derivative of each independent variable in Eq. (3.35) is calculated and Eq. (3.36) is obtained.

$$\begin{cases} \frac{\partial L}{\partial \omega} = 0 \Rightarrow \upsilon = \sum_{i=1}^{n} y_i\alpha_i x_i \\ \frac{\partial L}{\partial e} = 0 \Rightarrow \alpha_i = \gamma e_i \\ \frac{\partial L}{\partial b} = 0 \Rightarrow \sum_{i=1}^{n} y_i\alpha_i = 0 \\ \frac{\partial L}{\partial \alpha_i} = 0 \Rightarrow y_i\left[\upsilon^{\mathrm{T}}\phi(x_i + b)\right] = 1 - e_i \end{cases} \quad \text{To All} \quad i = 1, 2, \ldots n \tag{3.36}$$

Elimination method is applied to preliminary calculate Eq. (3.36), through which υ and e_i are eliminated. The calculated equation expressed in matrix form is demonstrated in Eq. (3.37).

$$\begin{cases} \begin{bmatrix} 0 & y^T \\ y & \mathbf{ZZ}^T + \gamma^{-1}\mathbf{I} \end{bmatrix}\begin{bmatrix} b \\ \alpha \end{bmatrix} = \begin{bmatrix} 0 \\ E_v \end{bmatrix} \\ \mathbf{y} = \left[y_1, y_2, \ldots, y_n\right]^T \\ \mathbf{Z} = [\phi(x_1), \phi(x_2), \ldots, \phi(x_n)]^T \\ \boldsymbol{\alpha} = [\alpha_1, \alpha_2, \ldots, \alpha_n]^T E_v[1, 1, \ldots 1]_{1\times v}^T \end{cases} \tag{3.37}$$

The inner product $\mathbf{ZZ}^T$ can be expressed by the kernel function $K(x_i, x_j)$ satisfying Mercer condition, as illustrated in Eq. (3.38).

$$\mathbf{ZZ}^T = y_i y_j K\left(x_i, x_j\right) \tag{3.38}$$

The classification decision function for Least Squares Support Vector Machine (LSSVM) can be represented by Eq. (1.36), and the choice of the kernel function can be made from Table 3.9. The Gaussian kernel is the most prevalent among the available kernel functions.

3.3.4 *Fault Feature Extraction Technology on the Basis of Frequency Domain Kurtosis Theory*

3.3.4.1 Frequency Domain Spectral Kurtosis Theory

Frequency domain kurtosis theory first proposed by Dwyer and Stewart is applied to detect the transient impact signal in the signal. Subsequently, Dwyer establishes a standardized approach to define the frequency domain kurtosis. Frequency domain kurtosis is the square of the expectation of the fourth-order central moment to the expectation of the second-order central moment.

Based on these studies, Antoni introduces the spectral kurtosis theory and employs it in the fault diagnosis of rotating machinery. The definition of spectral kurtosis theory is demonstrated in Eq. (3.39).

$$K_x(f) = \frac{S_4(f)}{(S_2(f))^2} - 2f \neq 0 \tag{3.39}$$

where $S_n(f) \triangleq E\{|X(f,t)|^n\}$, which indicates the n-th spectral moment of the signal, $E\{\cdot\}$ denotes mean operation, and $X(f,t)$ represents the complex envelope of signal $x(t)$ on frequency f.

3.3.4.2 Intelligent Fault Diagnosis Method for Wheelset Bearing

A wheelset-bearing fault diagnosis algorithm on the basis of cyclic correntropy and LSSVM is proposed in accordance with the above introduction, as demonstrated in Fig. 3.19. Besides, the algorithm proposed in this paper predominantly includes three parts: data acquisition, feature extraction and fault identification, the core is the fault feature extraction on the basis of the cyclic correntropy spectrum, and the specific steps are as follows:

Step 1: Set the bearing fault signal as $x[n]$, sampling frequency is F_s, time interval is N/F_s, number of overlapping samples is N_o, and the Length of each data block is N. Besides, the input signal is divided into L data blocks, and each data block has N samples.

Step 2: Utilize Silverman criterion to select the size of each data block core parameter σ_l $(l = 0, 1, 2, \ldots, L-1)$.

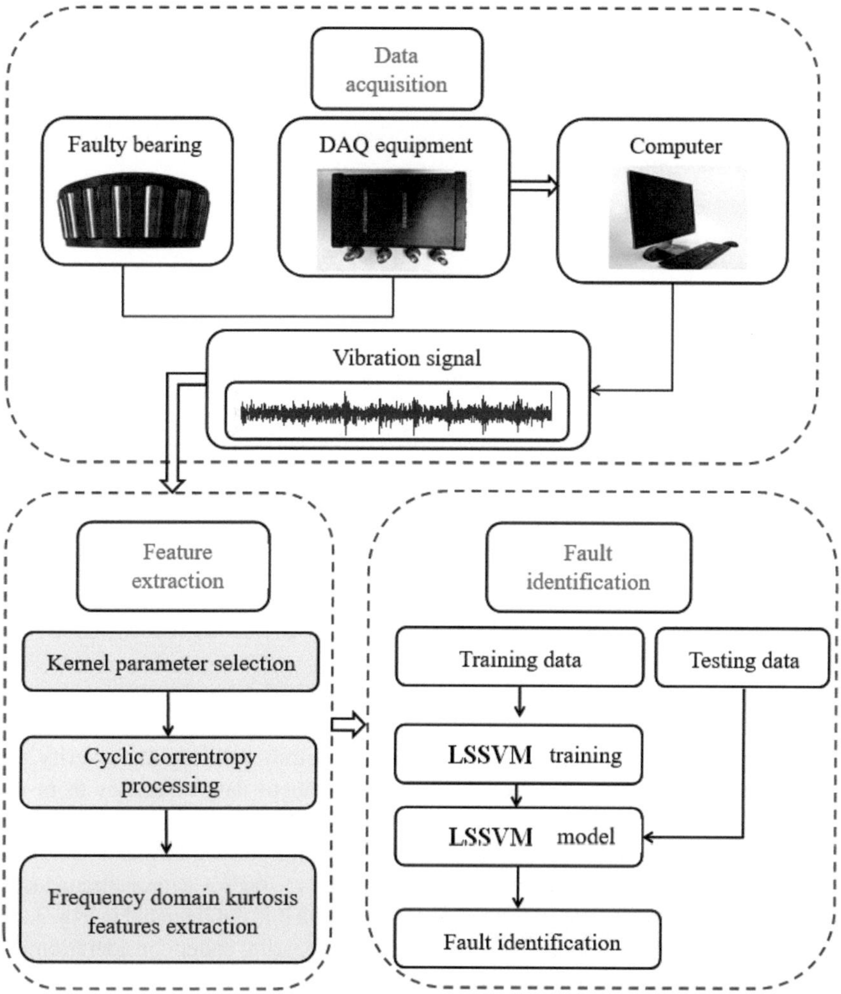

Fig. 3.19 Flowsheet of intelligent fault diagnosis method for wheelset bearing

Step 3: Calculate the average correntropy function of each data block, $l = 0, 1, 2, \ldots, L-1$.

$$M_l = \frac{1}{N^2} \sum_{\tau_n=0}^{N-1} \sum_{n=0}^{N-1} G_{\sigma_l}(x_l[n], x_l[n + \tau_n]) \tag{3.40}$$

Step 4: Calculate the frequency domain transformation of each data block as demonstrated in the formula. $\alpha[n] = n/N, n = 0, 1, 2, ...N-1, l = 0, 1, 2, ..., L-1$.

$$V_l^{\alpha_n}[\tau_n] = \sum_{n=0}^{N-1}\left\{\left[G_{\sigma_l}(x_l[n], x_l[n+\tau_n]) - M_l\right]e^{-j2\pi\alpha_n n}\right\} \tag{3.41}$$

Step 5: Compute the Fourier transform of each data block to acquire the cyclic correntropy spectrum.

$$T_l^{\alpha_n}[k] = \left|\frac{1}{N}\sum_{\tau_n=0}^{N-1} V_l^{\alpha_n}[\tau_n]e^{-j\frac{2\pi}{N}k\tau_n}\right| \tag{3.42}$$

Step 6: Map the cyclic correlation entropy spectrum to the cyclic frequency domain to acquire the cyclic frequency domain mapping spectrum of each data block.

Step 7: Extract the characteristic information of the signal on the basis of frequency domain spectral kurtosis of the signal acquired by the narrowband filtering principle.

Step 8: Partition the feature data into training data and test data and realize intelligent fault identification of wheelset bearing on the basis of LSSVM.

3.3.5 Simulation Analysis and Verification

3.3.5.1 Case Western Reserve University Data Validation

Two groups of bearing data are selected for algorithm verification to verify the effectiveness of the proposed method. The initial set of data originates from the bearing test bench at Case Western Reserve University in the United States, and the bearing test bench is demonstrated in Fig. 3.20. Moreover, the experimental platform includes an electric motor with two horsepower (1 hp = 746 W), a torque transducer, a dynamometer and electronic control equipment which is not revealed in Fig. 3.20, and the tested bearing supports the motor shaft. The bearing under consideration is a 6205-2RS JEM SKF deep groove ball bearing, where a single-point fault is introduced onto the bearing using electrical discharge machining technology. Specifically, fault data with a fault depth of 0.021 and a frequency of 12 kHz is chosen. The failure modes encompass scenarios of no failure (NF), rolling element failure (REF), inner ring failure (IRF), and outer ring failure (ORF).

On the basis of the calculation formula of fault frequency, the fault frequencies of bearing under distinct working conditions are demonstrated in Table 3.10. The selected bandwidths are 91–120 Hz, 121–150 Hz, and 151–180 Hz respectively. Based on the method proposed in this section, the data length is set to 6000, and the number of overlapping samples is $N_o = 2*N/3$. There are 58 samples corresponding to each failure mode, resulting in a total of 232 samples. 75% of the sample data is designated as the training dataset, while the remaining 25% is assigned as the test dataset. The identification of fault modes is carried out using the Least Squares Support Vector Machine (LSSVM).

Fig. 3.20 Bearing test bench in case Western Reserve University

Table 3.10 Theoretical fault frequency under distinct speeds

Motor speed (rpm)	ORF frequency (Hz)	REF frequency (Hz)	IRF frequency (Hz)
1730	103.36	135.91	156.14
1750	104.56	137.48	157.94
1772	105.87	139.21	159.93
1797	107.37	141.17	162.19

The fault identification results under distinct working conditions are demonstrated in Fig. 3.21. Taking Fig. 3.21a as an example to illustrate the experimental result, the NF data represented by the blue dots are wrongly divided into IRF data which is denoted by the black dots. The correct classification of REF data, IRF data, and ORF data affirms the effectiveness of the proposed method.

To verify the progressiveness of the method proposed in this research, two recently proposed fault identification methods are selected to make a comparison with the method proposed in this research. The first fault identification method, which is on the basis of the Teager–Kaiser (TK) energy factor, converts the vibration signal into the TK domain initially and then calculates the energy factor as the fault feature to determine the wheelset bearing fault. The effects of fault identification under distinct working conditions are demonstrated in Fig. 3.22.

The second fault identification method is on the basis of local mean decomposition (LMD). Feature extraction method on the basis of LMD, which is widely utilized in bearing fault identification, the method comprises three fundamental steps. In the first place, the input signal is decomposed into multiple PF components, and subsequently the sample entropy and energy ratio of PF components are calculated as the extracted features. Ultimately, the fault identification results are acquired by utilizing LSSVM. The experimental results are demonstrated in Fig. 3.23.

The experimental results illustrate that the method proposed in this research is superior to the comparison method. The result of fault identification on the basis of TK energy factor shows that several sample data are erroneously categorized

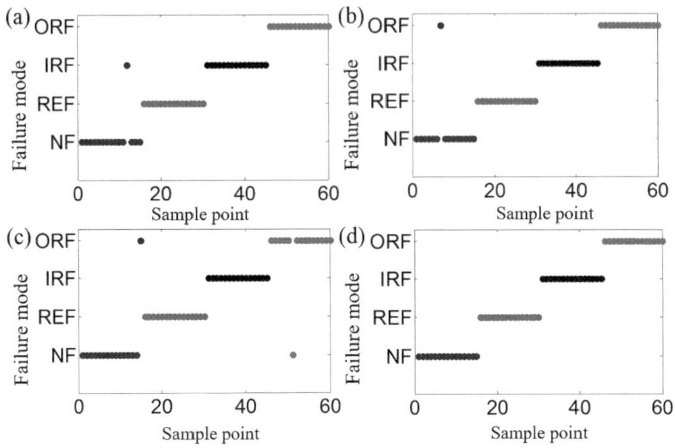

Fig. 3.21 Fault mode identification results at distinct speeds: **a** 1730 r/min; **b** 1750 r/min; **c** 1772 r/min; **d** 1797 r/min

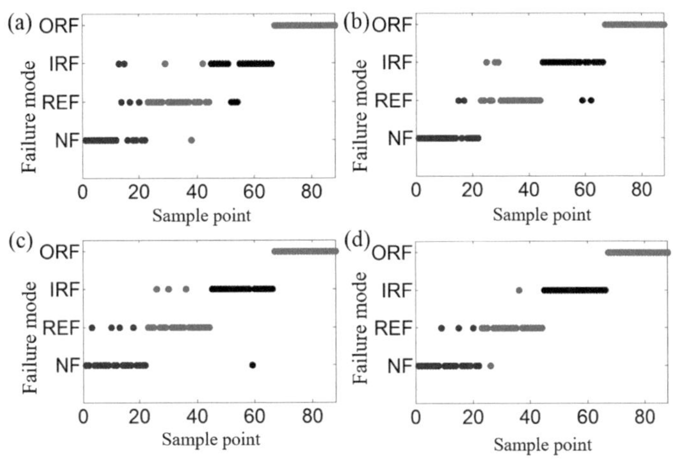

Fig. 3.22 Fault mode identification results on the basis of TK energy factor at distinct speeds: **a** 1730 r/min; **b** 1750 r/min; **c** 1772 r/min; **d** 1797 r/min

into different groups and this mistake becomes notably pronounced at 1730 r/min according to Fig. 3.23a. Despite the fact that the experimental accuracy of the feature extraction method on the basis of LMD is substantially enhanced, its classification accuracy remains not as accurate as the method proposed in this research. To further describe the fault identification effect of the method proposed in this research, the classification accuracy (CA), which is demonstrated in Eq. (3.43), is introduced to analyze the performance of the algorithm. The CA comparison results of distinct feature extraction methods are listed in Table 3.11. It is obvious that the classification effect of the method proposed in this research is obviously superior to other methods.

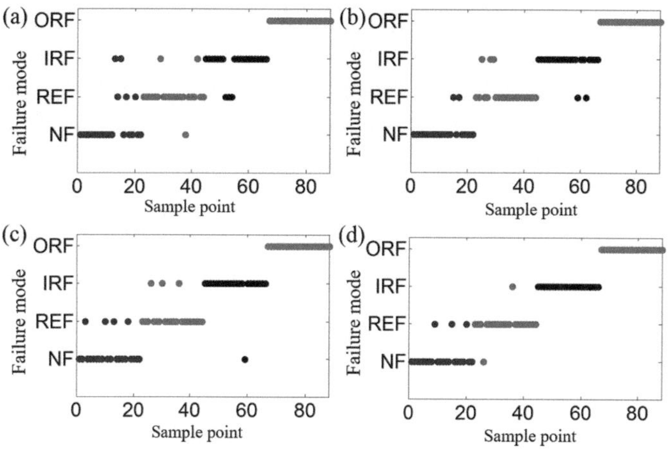

Fig. 3.23 Fault mode identification results on the basis of LMD at distinct speeds: **a** 1730 r/min; **b** 1750 r/min; **c** 1772 r/min; **d** 1797 r/min

$$CA = \frac{\text{Number of samples correctly classified}}{\text{Total number of samples}} \tag{3.43}$$

Table 3.11 CA comparison across different feature extraction methods

Motor speed (r/min)	Method	CA (%)				Average accuracy (%)
		NF	IRF	ORF	REF	
1730	TK	78.26	86.96	100	86.96	87.50
	LMD	100	86.67	100	100	98.33
	CCES	93.3	**100**	**100**	**100**	**98.33**
1750	TK	90.91	90.91	100	86.36	92.05
	LMD	100	93.33	93.33	100	93.33
	CCES	93.3	**100**	**100**	**100**	**98.33**
1772	TK	81.82	95.45	100	86.36	90.91
	LMD	100	86.67	86.67	100	96.67
	CCES	93.3	**100**	93.3	**100**	**96.67**
1797	TK	86.36	100	100	95.45	94.32
	LMD	100	93.33	100	100	98.33
	CCES	**100**	**100**	**100**	**100**	**100**

3.3.5.2 Data Analysis of Train Wheel Set Bearing

This experiment analyzes the fault data of the wheelset bearing of the train. During operation, the train will traverse turnouts and curves, generating impacts that transmit through the wheelset structure to reach the axle box. This gives rise to an impact noise environment in the wheelset axle box. To simulate the actual running environment of the train, the actual wheelset test platform is utilized to collect vibration data. Moreover, the test platform used to collect data on wheelset fault bearing is demonstrated in Fig. 3.24. The direct-current motor is connected to the main shaft through the belt transmission device, and its maximum speed can reach 1480 r/min. The wheelset bearing is positioned at each end of the main shaft, rotating in tandem with the main shaft. To emulate the vertical load and lateral impact induced by the train during operation, the hydraulic system regulates the transverse load device and the vertical load device. Additionally, a fan is employed to replicate the crosswind encountered by the train during its operation. Acceleration sensors are strategically placed in the 12-point direction (vertical load zone) and 3.point direction (perpendicular to the vertical load zone) of the wheelset bearing housing to capture data related to the wheelset fault bearing. The data is sampled at a frequency of 5120 Hz for accurate analysis. This setup ensures the coherence of the testing environment. Distinct rotational speeds and vertical loads are adopted to simulate the actual operation conditions of the train, and the transverse load generating impact noise is set as 20 kN. Additionally, the fault frequency obtained according to the theoretical calculation formula is demonstrated in Table 3.12.

In experimental group 1, the bandwidth selection is 1–50 Hz, 51–100 Hz, 101–150 Hz, while in experimental group 2, the bandwidth selection is 51–100 Hz, 101–150 Hz, 151–200 Hz, and the data length of each data block is 2560. According to the parameter selection method in the bearing data validation process of Case Western Reserve University, the number of samples for each failure mode is 65, and the total number of samples is 260. The identification of the failure mode is accomplished through the application of LSSVM.

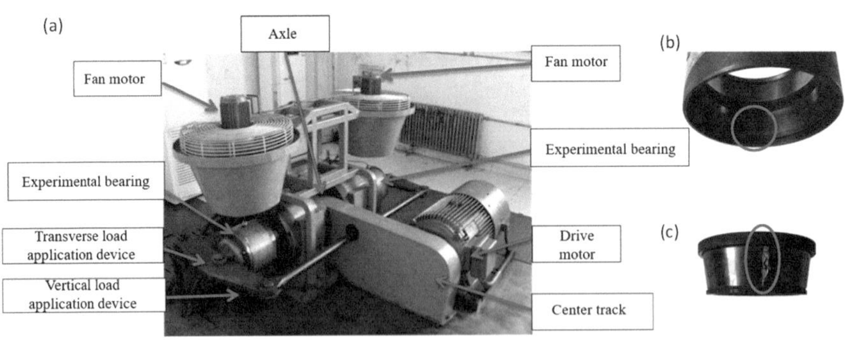

Fig. 3.24 Wheel set bearing test platform and fault bearing: **a** 1:1 wheel set bearing test platform **b** wheel set bearing ORF **c** wheel set bearing IRF

Table 3.12 Theoretical fault frequency of wheel set bearing

Grouping	Running speed (km/h)	Vertical load (kN)	ORF frequency (Hz)	REF frequency (Hz)	IRF frequency (Hz)
1	97	236	89.18	39.36	122.39
2	129	236	120.47	53.23	164.10
3	129	272	120.47	53.23	164.10

Figure 3.25 indicates the results of fault mode identification under distinct working conditions. Taking Fig. 3.25a as an example to illustrate the experimental results, it can be observed that several ORFs are incorrectly categorized as REF and NF, while other fault modes are accurately classified. The classification results illustrate the effectiveness of this method in dealing with fault signals under impulse noise interference.

Comparing the results of the fault identification method on the basis of TK energy factor and the results of the fault identification method on the basis of LMD with the results of the method in this research, as demonstrated in Figs. 3.26 and 3.27. Moreover, the experimental results also prove that the fault feature extraction method on the basis of LMD is superior to the fault feature extraction method on the basis of the TK energy factor. Additionally, the CA of the three methods is demonstrated in Table 3.13. Table 3.13 also indicates that the method proposed in this research is substantially superior to the other two approaches.

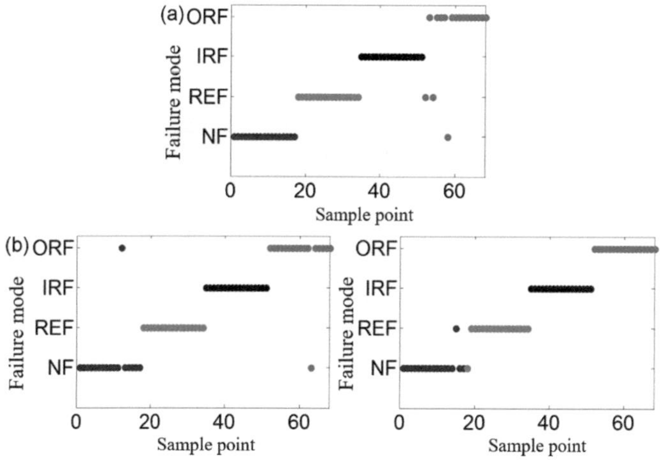

Fig. 3.25 Fault mode identification results under distinct working conditions: **a** Condition 1; **b** Condition 2; **c** Condition 3

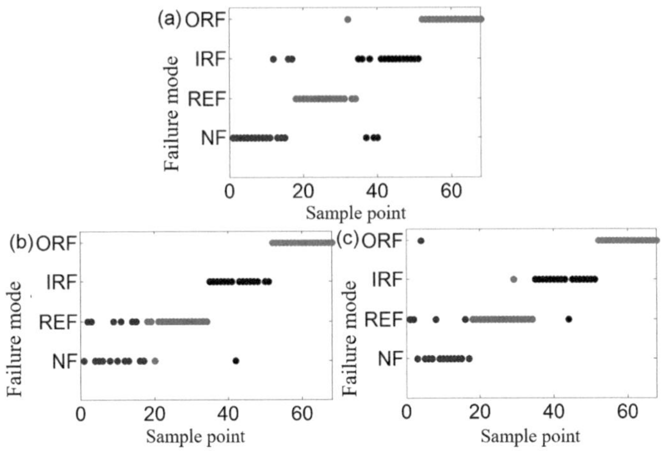

Fig. 3.26 Fault mode identification results on the basis of TK energy factor under distinct working conditions: **a** Condition 1; **b** Condition 2; **c** Condition 3

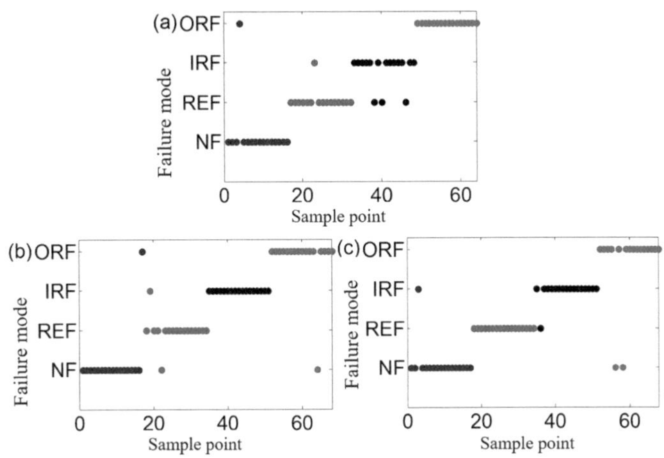

Fig. 3.27 Fault mode identification results on the basis of LMD under distinct working conditions: **a** Condition 1; **b** Condition 2; **c** Condition 3

3.4 Summary

This chapter presents a method for identifying the service state of key components of the train. Initially, an end-to-end component state identification method on the basis of a Convolution Neural Network (CNN) is proposed and a deep CNN is employed to determine the classification surface of the tensor domain, which is able to effectively avoid the defects of artificial design features, Leverage the information from

Table 3.13 CA of distinct methods

Running speed (km/h)	Vertical load (kN)	Method	CA (%)				Average accuracy (%)
			NF	IRF	ORF	REF	
97	236	TK	82.35	76.47	100	94.12	89.71
		LMD	93.75	87.5	100	93.75	92.19
		CCES	**100**	**100**	82.35	**100**	**95.59**
129	236	TK	64.71	94.12	100	94.12	85.29
		LMD	94.12	100	94.12	88.24	94.12
		CCES	**94.12**	**100**	94.12	**100**	**97.06**
129	272	TK	70.59	94.12	100	94.12	89.71
		LMD	94.12	94.12	88.24	100	94.12
		CCES	**94.12**	**100**	**100**	94.12	**97.06**

multiple sensors to its fullest extent, and realize automatic identification of mechanical components. Subsequently, utilizing the idea of joint distribution adaptation in transfer learning and combining it with an enhanced KNN classifier on the basis of correlation entropy distance measure, a migration diagnosis model with small samples is proposed in this chapter. This approach enables optimal utilization of experimental data within a laboratory setting and effectively addresses the challenge of limited samples, especially in cases with a scarcity of fault samples. Finally, an impulse noise suppression method for faulty bearings on the basis of cyclic correntropy is introduced. This method is on the basis of the mapping spectrum of cyclic correntropy spectrum in the cyclic frequency domain, extracts the kurtosis in the frequency domain as the fault feature, and employs LSSVM to identify the intelligent fault of bearings. The experimental results indicate that the proposed method is progressive and effective.

References

1. Li, Y., Liang, X., Lin, J., et al. (2018). Train axle bearing fault detection using a feature selection scheme based multi-scale morphological filter. *Mechanical Systems & Signal Processing, 101*, 435–448.
2. Sharma, V., & Parey, A. (2017). Frequency domain averaging based experimental evaluation of gear fault without tachometer for fluctuating speed conditions. *Mechanical Systems & Signal Processing, 85*, 278–295.
3. Mitchell, T. M. (1997). *Machine learning*. McGraw-Hill.
4. Sun, S., & He, G. (2015). Fault diagnosis method of rolling bearing based on dual-tree rational-dilation complex wavelet packet transform and SVM. *Computer Science, 42*, 140–143.
5. Liang, Y., Cao, J., & Ji, X. (2019). Fault diagnosis of rolling bearing based on EEMD and CS-SVM. *Journal of Mechanical & Electrical Engineering, 36*, 622–627.
6. Chen, F., Li, M., & Chen, B. (2016). Fault diagnosis of rolling bearing based on hybrid feature set and weighted KNN. *Journal of Mechanical Transmission, 040*, 138–143.

7. Pandya, D. H., Upadhyay, S. H., & Harsha, S. P. J. (2013). Fault diagnosis of rolling element bearing with intrinsic mode function of acoustic emission data using APF-KNN. *2Expert Systems with Applications, 40*, 4137–4145.
8. Zhuang, F., Luo, P., et al. (2015). Survey on transfer learning research. *Journal of Software, 26*, 26–39.
9. Pan, S. J., & Qiang, Y. A. (2009). Survey on transfer learning. *IEEE Educational Activities Department, 22*(10), 1345–1359.
10. Pan, S. J., Tsang, I. W., & Kwok, J. T., et al. (2011) Domain adaptation via transfer component analysis. *IEEE Transactions on Neural Networks, 22*, 199–210.
11. Feng, H., Hu, M., & Yang, G. (2019). Tree species recognition based on overall tree image and ensemble of transfer learning. *Transactions of the Chinese Society for Agricultural Machinery, 050*(235–242), 279.
12. Lecun, Y., Bengio, Y., & Hinton, G. (2015). Deep learning. *Nature, 521*, 436.
13. Li, D., Li, J., & Huang, J. T., et al. (2011) Recent advances in deep learning for speech research at Microsoft. In *Proceedings of IEEE International Conference on Acoustics,Prague,Czech Republic*.
14. Shih, Y. F., Yeh, Y. M., & Lin, Y. Y., et al. (1997). Deep co-occurrence feature learning for visual object recognition. In *Proceedings of IEEE Conference on Computer Vision & Pattern Recognition*. San Juan, USA.
15. Wilkinson, T., & Brun, A. (2015). A novel word segmentation method based on object detection and deep learning. ISCV 2015, Las Vegas, USA.
16. Zhu, H., Rui, T., & Wang, X., et al. (2015) Fault diagnosis of hydraulic pump based on stacked autoencoders. In *Proceedings of IEEE International Conference on Electronic Measurement & Instruments*. Qingdao.
17. Amirolad, A., Arashloo, S. R., & Amirani, M. C. (2016) Multi-layer local energy patterns for texture representation and classification. *Visual Computer*, 1633–1644.
18. Gawehn, E., Hiss, J. A., & Schneider, G. (2016). Deep learning in drug discovery. *Molecular Informatics, 35*, 3–14.
19. Yongjin, P., & Manolis, K. (2015). Deep learning for regulatory genomics. *Nature Biotechnology, 33*, 825–826.
20. Chen, Z., & Li, W. (2017). Multisensor feature fusion for bearing fault diagnosis using sparse autoencoder and deep belief network. *IEEE Transactions on Instrumentation & Measurement, 66*, 1–10.
21. Mao, W., He, J., & Li, Y., et al. (2016). Bearing fault diagnosis with auto-encoder extreme learning machine: A comparative study. *Archive Proceedings of the Institution of Mechanical Engineers Part C Journal of Mechanical Engineering Science 1989–1996, 203–210*, 231.
22. Xia, M., Li, T., Liu, L., et al. (2017). An intelligent fault diagnosis approach with unsupervised feature learning by stacked denoising autoencoder. *Iet Science Measurement & Technology, 11*, 687–695.
23. Feng, J. (2016). Deep neural networks: A promising tool for fault characteristic mining and intelligent diagnosis of rotating machinery with massive data. *Mechanical Systems & Signal Processing, 72–73*, 303–315.
24. Guo, L., Gao, H., Huang, H., et al. (2016). Multifeatures fusion and nonlinear dimension reduction for intelligent bearing condition monitoring. *Shock & Vibration, 2016*, 1–10.
25. Verma, N. K., Gupta, V. K., & Sharma, M., et al. (2013) Intelligent condition based monitoring of rotating machines using sparse auto-encoders. In *Proceedings of Prognostics & Health Management*. Gaithersburg, MD.
26. Feng, J., Lei, Y., Guo, L., et al. (2017). A neural network constructed by deep learning technique and its application to intelligent fault diagnosis of machines. *Neurocomputing, 272*, 619–628.
27. Shao, H., Jiang, H., Zhao, H., et al. (2017). A novel deep autoencoder feature learning method for rotating machinery fault diagnosis. *Mechanical Systems & Signal Processing, 95*, 187–204.
28. Sang, J. L., Yun, J. P., Koo, G., et al. (2017). End-to-end recognition of slab identification numbers using a deep convolutional neural network. *Knowledge-Based Systems, 132*, 718–721.

29. Lawrence, S., Giles, C. L., Tsoi, A. C., et al. (1997). Face recognition: A convolutional neural-network approach. *IEEE Transactions on Neural Networks, 8*, 98–113.
30. Shou, Z., Wang, D., Chang, S. F. (2016). Temporal action localization in untrimmed videos via multi-stage CNNs. In *Proceedings of the IEEE Conference on Computer Vision and Pattern Recognition*. Las, Vegas
31. Pellegrini, T. (2017). Densely connected CNNs for bird audio detection. In *Proceedings of Signal Processing Conference*. Kos, Greece.
32. Zhou H. Acoustic diagnosis system for early fault of rolling bearing of railway train. Dual Use Technologies & Products, 2016, 41–41.
33. Chen D. Principle and application of trackside acoustic diagnosis system for early failure of freight car rolling bearing. Shanghai Railway Science & Technology, 2011, 107–108.
34. Li B., Liu R., Zhang J. Trackside acoustic diagnosis system for early failure of freight car rolling bearing. Chinese Railways, 2006, 35–38.
35. Zhang, J., Zhao, C., Chen, S., & Zhang, Y. (2015). Research on trackside acoustic diagnosis system for early fault of rolling bearings of railway passenger cars. *Rolling Stock, 53*, 29–32.
36. Ren, L., Cui, J., Sun, Y., et al. (2017). Multi-bearing remaining useful life collaborative prediction: A deep learning approach. *Journal of Manufacturing Systems, 43*, 248–256.
37. Soualhi, A., Razik, H., Clerc, G., et al. (2014). Prognosis of bearing failures using hidden Markov models and the adaptive neuro-fuzzy inference system. *IEEE Transactions on Industrial Electronics, 61*, 2864–2874.
38. Hong, S., Zhou, Z., Zio, E., et al. (2014). Condition assessment for the performance degradation of bearing based on a combinatorial feature extraction method. *Digital Signal Processing, 27*, 159–166.
39. Liao, L., & Lee, J. (2009). A novel method for machine performance degradation assessment based on fixed cycle features test. *Journal of Sound and Vibration, 326*, 894–908.
40. Hong, S., Zhou, Z., Zio, E., et al. (2014). An adaptive method for health trend prediction of rotating bearings. *Digital Signal Processing, 35*, 117–123.
41. Lei, Y., Zuo, M. J., He, Z., et al. (2010). A multidimensional hybrid intelligent method for gear fault diagnosis. *Expert Systems with Applications, 37*, 1419–1430.
42. Qu, J., Zuo, M. J. (2012). An LSSVR-based algorithm for online system condition prognostics. *Expert Systems with Applications, 39*, 6089–6102.
43. Peng, C., Wang, X., Zhang, Z., et al. (2015). Moors-kurtogram based bearing fault diagnostic methodfor high speed trains. *Manufacturing Automation, 37*, 41–43.
44. Huang, H., Gao, H., Li, D., et al. (2017). Study on evaluation of incipient performance degradation of rolling bearing. *Mechanical Science and Technology for Aerospace Engineering, 36*, 1771–1777.
45. Tamilselvan, P., & Wang, P. (2013). Failure diagnosis using deep belief learning based health state classification. *Reliability Engineering & System Safety, 115*, 124–135.
46. AlThobiani, F., & Ball, A. (2014). An approach to fault diagnosis of reciprocating compressor valves using Teager-Kaiser energy operator and deep belief networks. *Expert Systems with Applications, 41*, 4113–4122.
47. Tang, B., Gong, X.-J., Wei, W., et al. (2017). Intelligent fault diagnosis of the high-speed train with big data based on deep neural networks. *IEEE Transactions on Industrial Informatics, 13*, 2106–2116.
48. Zhao, Y., Guo, Z. H., & Yan, J. M. (2017). Vibration signal analysis and fault diagnosis of bogies of the high-speed train based on deep neural networks. *Journal of Vibroengineering, 19*, 2456–2474.
49. Shao, H., Jiang, H., Zhang, X., & Niu, M. (2015). Rolling bearing fault diagnosis using an optimization deep belief network. *Measurement Science and Technology, 26*, Article 115002.
50. Zhang, W., Li, C., Peng, G., et al. (2018). A deep convolutional neural network with new training methods for bearing fault diagnosis under noisy environment and different working load. *Mechanical Systems and Signal Processing, 100*, 439–453.
51. Niknam, S. A., Songmene, V., & Au, Y. H. J. (2013). The use of acoustic emission information to distinguish between dry and lubricated rolling element bearings in low-speed rotating machines. *International Journal of Advanced Manufacturing Technology, 69*, 2679–2689.

52. Frankel, F., & Reid, R. (2008). Big data: Distilling meaning from data. *Nature, 455*, 30.
53. Report, I. C. (1985). Motor Reliability Working Group. Report on large motor reliability survey of industrial and commercial installations. *IEEE Transactions on Industry Applications, 4*, 865–872.
54. Website, T C W.R.U.B.D.C. Bearing Data Center Seeded Fault Test Data [EB/OL]. https://engineering.case.edu/bearingdatacenter/download-data-file.
55. Zhang, W., Peng, G., Li, C., et al. (2017). A New Deep Learning Model for Fault Diagnosis with Good Anti-Noise and Domain Adaptation Ability on Raw Vibration Signals. *Sensors, 17*, 425.
56. Brownlee, J. How To Improve Deep Learning Performance [EB/OL]. https://www.freecodecamp.org/news/improve-image-recognition-model-accuracy-with-these-hacks/.

Chapter 4
Real-Time Reliability Analysis of Key Components of Rail Trains

4.1 Introduction

With the continuous enhancement of design and manufacturing technology, as well as the materials used, the operational lifespan of mechanical devices is steadily increasing, particularly within the realm of rail transit. For instance, the design service life of the bearing on the bogie axle box is over 20 years. These products usually have only a small number of failures or no failures in life tests including accelerated life test methods. Nonetheless, the traditional reliability evaluation methods including fault tree analysis (FTA) [1], event tree analysis (ETA) [2], failure mode effects and criticality analysis (FMECA) [3], etc., predominantly focus on the "average" characteristics of the same kind of product with fault record data insurmountable. There is a gap between these methods in describing the performance degradation of specific components. Moreover, devices of the same have distinct degradation trajectories under distinct operating environments [4], and even devices of the same type with the same operating environments may also have distinct degradation trajectories, arising from randomness in manufacturing process or other unexpected factors. Since real-time reliability evaluation has become increasingly essential nowadays, performance monitoring data is typically employed for reliability evaluation of the specific product, instead of traditional failure records. The rapid advancement in sensor technology has significantly augmented the availability of data for service performance evaluation and reliability analysis. This development greatly enhances the potential for accurate reliability assessment and prediction.

The concept of reliability analysis on the basis of performance degradation data was first proposed by Gertsbackh and Kordonskiy in 1969 [5]. Subsequently, numerous experts and scholars have initiated exploratory research in this domain. Currently, it can be predominantly divided into two kinds of performance degradation reliability analysis methods on the basis of regression model and stochastic process. A simple regression model is employed in escalators Avante model on the basis of time series analysis [6]. The effect of variable amplitude service loading

Y. Qin et al., *Intelligent Safety and Reliability Analysis of Rail Transit Trains*,
https://doi.org/10.1007/978-981-97-4170-0_4

conditions and wear phenomenon are investigated on the basis of nonlinear regression reliability analysis for rail steel [7]. Reference [8] integrated the fault data with the state detection data of performance degradation, and employed the DTS (Degradation Threshold Shock) method to analyze the reliability of competitive failure phenomenon. The frequently employed stochastic processes are predominantly Wiener processes [9, 10], Gamma process [11], Poisson process [12], Markov process [13], and etc. Some reliability analysis methods typically assume that the device only has two states (normal and failure) [14–16]. The potential changes during device degradation had been ignored, which made the reliability results too crude to guide practical application. In comparison with two-state (normal, fault) reliability analysis models, the multistate model is more suitable for actual equipment operation. Reference [17] provided a reliability evaluation framework on the basis of multi-performance conversion, which is conducted on the conversion capabilities of the components. A multistate physics model is introduced for dynamic reliability evaluation, incorporating crack propagation state transitions. The Forman model is specifically applied to analyze structures with multiple cracks [18]. As a powerful tool for multistate performance degradation modeling, the Markov process of distinct types has obtained numerous achievements, including the Discrete Time Markov Process (DTMP) [19, 20], Continuous Time Markov Process (CTMP) [21, 22], Discrete-Time Hidden Aging Markov Process (DTHAMP) [23, 24], nonhomogeneous continuous-time hidden Markov process (NHCTHMP) [25, 26], homogeneous discrete-time hidden semi-Markov process (HDTHSMP) [27–29], etc. Nonetheless, the number of device states was set directly, which may not objectively reflect the overall trend of the device's performance, and may lead to a decrease in model accuracy. Very few literature have studied the actual number of device's states. Reference [30] achieved the optimal number of states by multiple trials under distinct state numbers with BIC and AIC as evaluation indicators. Nonetheless, the time-consuming problem stands out with the increase in available data. Besides that, the state transition matrix of most Markov models is fixed, which means no matter how the device performance is degraded, the device reliability does not change, when it is detected in a certain state of such a Markov model. The longer the system or device stays in a certain state, the greater the probability of transitioning out of the state, and accordingly the mentioned Markov model cannot reflect the performance degradation within the same state interval over time which is not in line with reality [31], which is a remarkable obstacle in dealing with time-varying characteristics of device performance. Moreover, the majority of degradation reliability evaluation methods, particularly for mechanical devices, are on the basis of single monitoring data, including vibration data. Multi-source monitoring is the trend of device performance condition monitoring, and more information is provided for accurate detection and prediction [32]. Effectively harnessing multi-source data is an urgent issue that requires resolution.

Hence, in this research, a real-time reliability evaluation model on the basis of automatic partition method and time-varying Markov chain is proposed, in which the automatic partition is utilized to select the state number of Markov process scientifically, while the time-varying properties of device performance are described by the time-varying transition matrix of Markov process. The rest of this research is structured as follows: the basic concept of Tsallis entropy working as data feature is presented in the second section first. Subsequently, the framework of the proposed real-time reliability evaluation algorithm on the basis of the automatic partition time-varying Markov process is illustrated. An experiment is conducted in Sect. 4.3 to demonstrate the superiority of the proposed algorithm. Specifically, the study is conducted on bearings using condition monitoring vibration signals in two directions and temperature signals. The findings of the article are summarized in the concluding section.

4.2 Materials and Methods

4.2.1 Tsallis Entropy

Tsallis Entropy is non-extensive entropy, it is an expression of entropy with exponent q proposed by Brazilian physicist Constantino Tsallis in 1988 [33]. Its mechanism is on the basis of the non-extensive statistics derived from Boltzmann–Gibbs theory. The Tsallis Entropy demonstrates excellent performance in describing systems with long-range interactions, long-term memory effects, or systems evolving in a multifractal-like space–time. These characteristics make the Tsallis Entropy widely applicable in fields like physics, medicine, and economics. The Tsallis entropy is concave, stable, and limited in the rate of entropy generation per unit time [34]. Despite the fact that Shannon entropy, Renyi entropy, etc., have extremely rich applications in the field of bearing analysis, these entropies are not excellent enough to describe the bearing performance degradation with vibration signal. The comparison has been demonstrated in Fig. 4.3. Consequently, in this research, the Tsallis entropy is extracted as a data feature from the vibration signal. The definition is provided below

$$H_q^T = \frac{1}{q-1}\left(1 - \sum_{i=1}^{n} p_i^q\right) \tag{4.1}$$

where q describes the non-extension degree of target object. In this research, it indicates the long-range correlation of time series data. n represents the total data number in one sample. p_i indicates the i-th value of sample.

The purpose of Tsallis entropy is to determine a clear q value (generally not equal to 1) to comprehensively characterize phenomena that exhibit neither regularity nor complete chaos or randomness. Following numerous experiments, we found that $q = 2$ can better describe the changing characteristics of bearing.

4.2.2 The Automatic Partition Time-Varying Markov Model for Reliability Evaluation

The reliability analysis method on the basis of automatic partition time-varying Markov process predominantly includes five steps: Degradation feature extraction, state number decision, training data preparation, time-varying Markov process modeling, and reliability assessment. The automatic partition method is employed to autonomously segment the sample data related to component degradation into multiple stages. This helps circumvent the need for manual and time-consuming selection of the number of Markov process states. In the data processing procedure, to deal with the distinct degradation processes of bearings, the slope-based dynamic time warping method is employed to effectively regularize the collected samples, so as to get a more accurate bearing Markov model. On the basic Markov process, a time-varying transition probability matrix is proposed to describe the changing trend of the device's own characteristics with time, which offers an essential approach for a more precise assessment of device state reliability and life prediction. The functions for reliability and remaining useful life are provided in the last step.

The basic steps are provided as follows and demonstrated in Fig. 4.1.

(1) Extract feature on the full-life running vibration data.
(2) The automatic partition method is employed to decide the state number of the Markov model, corresponding to the degradation state of the device.
(3) Prepare the training data. In this step, the slope-based dynamic time warping method is applied for sample time warping, completing the processing of the sample data.
(4) Establish the time-varying Markov model. The model parameters are re-estimated utilizing the conditional probability formula for calculating the initial forward probability, backward probability, and state sequence.
(5) Reliability and remaining useful life calculation and prediction. Determining the current state, and getting the sojourn time in the current state i. Subsequently, calculating it in the remaining states $i+1$, $i+2$, ..., N. Calculating the reliability and remaining useful life of the equipment with current state i and sojourn time st_t^i.

4.2.2.1 The Automatic Partition Method for State Number Determination of Markov Model

The automatic partition method on the basis of fuzzy time series segmentation is used as to automatically obtain the state number of Markov model.

Assuming x_l indicates a M dimensional multivariate sample, and S is the sample sequence with length L, where $S = \left\{d_l = [t_l, x_l^T]^T | 1 \le l \le L\right\}$, $x_l = \{x_{lm} | m = 1, 2, ..., M\}$. According to the trend of data change, that is on the basis

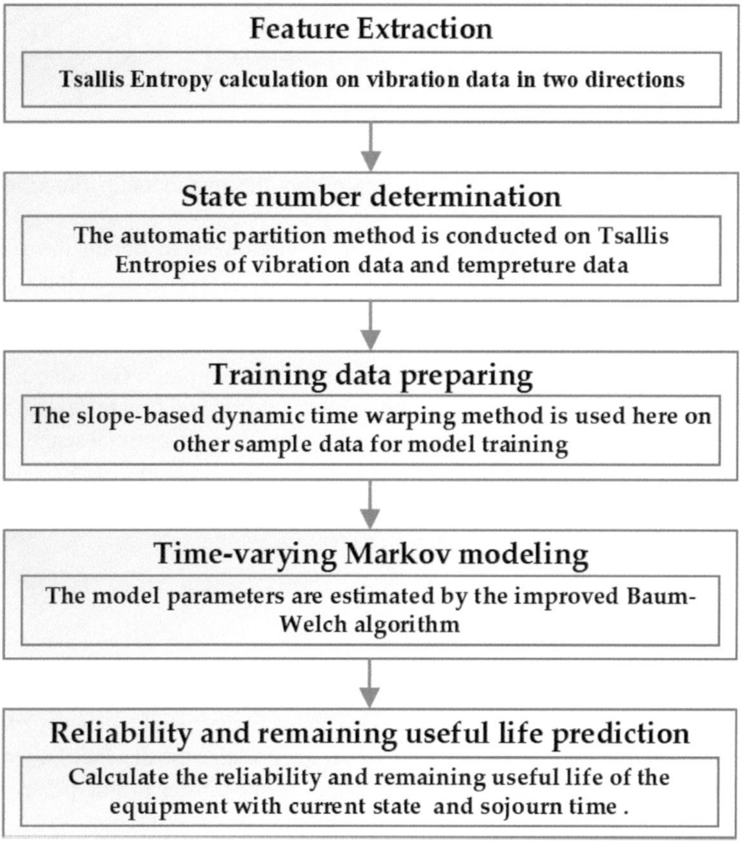

Fig. 4.1 Basic steps of the proposed method

of the equipment performance degradation, sample sequence S can be partition into several segments.

In this research sample sequence S is assumed to be divided into P segments, noted as $\mathrm{Sub}_S^P = \left\{\mathrm{Sub}_S^p | p = 1, 2, ..., P\right\}$, and the boundaries are $s_1 < s_2 < \cdots s_P$, where $s_1, s_2, ..., s_P \in S$, $s_1 = d_1$, $s_P = d_L$. That is $\mathrm{Sub}_S^p = \left(s_{p-1} + 1, s_p\right), p = 1, 2, ..., P$.

Moreover, we define the sample sequence segmentation objective function as $C\left(\mathrm{Sub}_S^P\right)$.

$$\mathrm{C}\left(\mathrm{Sub}_\mathrm{S}^\mathrm{P}\right) = \sum_{\mathrm{p}=1}^{\mathrm{P}} \mathrm{C}\left(\mathrm{Sub}_\mathrm{S}^\mathrm{p}\right) = \sum_{\mathrm{p}=1}^{\mathrm{P}} \mathrm{C}\left(\mathrm{s}_{\mathrm{p}-1} + 1, \mathrm{s}_\mathrm{p}\right) \tag{4.2}$$

The objective function is generally represented as the distances between the monitoring data in each segment and their corresponding fitting function.

$$C\left(Sub_S^P\right) = \sum_{p=1}^{P} \sum_{l=s_{p-1}+1}^{s_p} \left\| x_l - v_p^x \right\|^2 = \sum_{p=1}^{P} \sum_{l=1}^{L} \beta_p(t_l) D^2\left(x_l - v_p^x\right) \tag{4.3}$$

where v_p^x indicates the cluster center of the $p-$ th segment, $D^2\left(x_l, v_p^x\right)$ implies the distance between x_l and v_p^x, $\beta_p(t_l)$ represents the membership function of x_l subordinate to the $p-$ th segment.

The objective function mentioned above is minimized to obtain the segment boundaries with optimal segment positions $s_{p-1}+1, s_p, p = 1, 2, ..., P$. It is typically conducted by numerous heuristic algorithms.

$\beta_p(t_l)$ is defined as a (0,1) binary function in references [35, 36], which is not appropriate in practice. In this research, the multivariate mixture Gaussian function is introduced into cluster prototype as the sequence fitting function to describe fuzzy characteristic of boundaries between segments. Moreover, the optimal segment positions are obtained by minimizing the weighted sum of squared distances between the monitoring data in each segment and the center of the cluster prototype.

$$C\left(Sub_S^P\right) = \sum_{p=1}^{P} \sum_{l=1}^{L} \left(\mu_{p,l}\right)^m D^2\left(d_l, \eta_p\right) \tag{4.4}$$

where $\mu_{p,l}$ indicates the membership of $d_l = \left[t_l, x_l^T\right]^T$ belonging to the $p-$ th segment, $p = 1, 2, ..., P$; $m = [1, \infty)$ demonstrates the weighted index of fuzzy clustering, generally take $m = 2$ [37]; $D^2\left(d_l, \eta_p\right)$ represents the distance between the monitoring data in each segment and the center of the cluster prototype; η_p shows the clustering prototype function of $p-$ th segment.

Assuming the expectation and the covariance matrix of the used multivariate Gaussian distribution for sample sequence are v_p and F_p, respectively. The probability density function $f(d_l|\eta)$ denoting the adscription of monitoring data to the $p-$ th segment is

$$f(d_l|\eta) = \sum_{p=1}^{P} f\left(d_l|\eta_p\right) f\left(\eta_p\right) \tag{4.5}$$

Where, $f\left(d|\eta_p\right) = \frac{1}{(2\pi)^{(n+1)/2}\sqrt{\det\left(F_p\right)}} \exp\left(-\frac{1}{2}\left(d - v_p\right)^T \left(F_p\right)^{-1}\left(d - v_p\right)\right)$; $f\left(\eta_p\right)$ is the unconditional clustering probability, $\sum_{p=1}^{P} f\left(\eta_p\right) = 1$; η_p is the clustering prototype function of the $p-$ th segment, $\eta_p = \left\{f\left(\eta_p\right), v_p, F_p | p = 1, 2, ..., P\right\}$.

According to probability theory, the distance function and the membership in the $p-$ th segment are inversely proportional on the basis of the Gath–Geva clustering algorithm [38]. The time and feature variables of sample sequence are independent with each other. Subsequently,

$$
\begin{aligned}
f\left(d_l|\eta_p\right) &= \frac{1}{D^2\left(d_l,\eta_p\right)} \\
&= \underbrace{\alpha_p}_{p(\eta_p)} \underbrace{\frac{1}{\sqrt{2\pi\sigma_{p,t}^2}}\exp\left(-\frac{1}{2}\frac{\left(t_l-v_p^t\right)^2}{\sigma_{p,t}^2}\right)}_{p(t_l|\eta_p)} \\
&\times \underbrace{\frac{1}{(2\pi)^{r/2}\sqrt{\det\left(A_p\right)}}\exp\left(-\frac{1}{2}\left(x_l-v_p^x\right)^T\left(A_p\right)^{-1}\left(x_l-v_p^x\right)\right)}_{p(x_l|\eta_p)}
\end{aligned}
\tag{4.6}
$$

where α_k is the initial probability of clustering, $\frac{1}{\sqrt{2\pi\sigma_{p,t}^2}}\exp\left(-\frac{1}{2}\frac{\left(t_l-v_p^t\right)^2}{\sigma_{p,t}^2}\right)$ is the distance between the time variable and the clustering center of time variable, $\frac{1}{(2\pi)^{r/2}\sqrt{\det\left(A_p\right)}}\exp\left(-\frac{1}{2}\left(x_l-v_p^x\right)^T\left(A_p\right)^{-1}\left(x_l-v_p^x\right)\right)$ is the distance between the feature variable in the $l-$ th sample and the feature cluster center, v_p^x is the feature cluster center of $p-$ th segment, r is the rank of the feature variable distance norm A_p.

There are many ways to calculate the distance norm A_p. The Mahalanobis distance is a suitable option and can be employed to modify the correlation between variables. Let $A_p = F_p$, where F_p is the fuzzy covariance matrix of the multivariate Gaussian distribution,

$$
F_p = \frac{\Sigma_{l=1}^{L}\left(\mu_{p,l}\right)^m\left(\left(x_l-v_p^X\right)\left(x_l-v_p^X\right)^T\right)^T}{\Sigma_{l=1}^{L}\left(\mu_{p,l}\right)^m} \tag{4.7}
$$

Principal Component Analysis (PCA) is introduced to eliminate the strong correlations between variables while dimensionality reduction.

Assuming $\lambda_{p,q}$, $q = 1, 2, ..., Q$ are the Q non-zero eigenvalues of F_p in descending order. Their corresponding eigenvectors are $u_{p,q}$, $q = 1, 2, ..., Q$. It can be obtained that

$$
F_p = U_p \Lambda_p U_p^T \tag{4.8}
$$

$$
\Lambda_p = \begin{vmatrix} \lambda_{p,1} & & & \\ & \lambda_{p,2} & & \\ & & \ddots & \\ & & & \lambda_{p,Q} \end{vmatrix} \tag{4.9}
$$

$$
U_p = \left[u_{p,1}, u_{p,2}, \ldots, u_{p,Q}\right] \tag{4.10}
$$

The feature variables of the sample sequence are reduced to Q-dimension by PCA method, and $y_{p,l} = W_p^{-1}(x_l) = W_p^T(x_l)$ is obtained, where $W_p = U_{p,Q}\Lambda_{p,Q}^{1/2}$.

According to probabilistic principal component analysis [39], $W_p = U_{p,Q}\left(\Lambda_{p,Q} - \sigma_{p,x}^2 I\right)^{1/2} R_{pS}$, where R_p is a $Q \times Q$ orthogonal transformation matrix, $\sigma_{p,x}^2 = \frac{1}{M-Q}\sum_{j=Q+1}^{M} \lambda_{p,q}$.

Until now, the automatic partitioning algorithm of fuzzy segmentation has been transformed into an operational research optimization problem. The model is demonstrated below

$$C\left(Sub_S^P\right) = \sum_{p=1}^{P}\sum_{l=1}^{L}\left(\mu_{p,l}\right)^m D^2\left(d_l, \eta_p\right) \tag{4.11}$$

$$\begin{aligned} & U = \left[\mu_{p,l}\right]_{P\times L,\mu_{p,l}} \in [0, 1], \forall p, l \\ & \sum_{p=1}^{P} \mu_{p,l} = 1, \ldots\ldots\ldots\forall l \\ & 0 < \sum_{l=1}^{L} \mu_{p,l} < L, \cdots \forall p \end{aligned} \tag{4.12}$$

The alternate optimization algorithm and the Picard iterative algorithm can be used to solve this problem.

The basic steps are as follows:

(1) Initialization

The initial segments number of S and the dimension of the eigenvector space are given at first. Then select the appropriate stopping condition ε, $\varepsilon > 0$ and initialize W_p, v_p^x, $\sigma_{p,x}^2$, $\mu_{p,l}$;

(2) Loop Calculation.

(i) Calculate the cluster prototype η_p,

Cluster initial probability:

$$\alpha_p = \frac{1}{L}\Sigma_{l=1}^{L}\mu_{p,l} \tag{4.13}$$

Cluster center:

$$v_p^X = \frac{\sum_{l=1}^{L}\left(\mu_{p,l}\right)^m\left(x_l - W_p\langle y_{p,l}\rangle\right)}{\Sigma_{l=1}^{L}\left(\mu_{p,l}\right)^m} \tag{4.14}$$

where $y_{p,l} = M_p^{-1} W_p^T \left(x_l - v_p^x \right)$, $M_p = \sigma_{p,x}^2 I + W_p^T W_p$;

Weight update:

$$\tilde{W}_p = F_p W_p \left(\sigma_{p,x}^2 I + M_p^{-1} W_p^T F_p W_p \right)^{-1} \tag{4.15}$$

Variance update:

$$\sigma_{p,x}^2 = \frac{1}{Q} tr \left(F_p - F_p W_p M_p^{-1} W_p^T \right) \tag{4.16}$$

Distance norm:

$$A_p = \sigma_{p,x}^2 I + \tilde{W}_p \tilde{W}_p^T \tag{4.17}$$

Calculation of cluster prototype center and standard deviation parameter of time parameter for sample sequence:

$$v_p^t = \frac{\Sigma_{l=1}^L \left(\mu_{p,l} \right)^m t_l}{\sum_{l=1}^L \left(\mu_{p,l} \right)^m}, \cdots \sigma_{p,t}^2 = \frac{\Sigma_{l=1}^L \left(\mu_{p,l} \right)^m \left(t_l - v_p^t \right)^2}{\sum_{l=1}^L \left(\mu_{p,l} \right)^m} \tag{4.18}$$

(ii) Cluster merging

The similarity factor $S_{PCA}^{a,b}$ on the basis of PCA proposed by Krzanowski is applied as one of the merging criteria for two adjacent segments Sub_S^a, Sub_S^b that whether they need to be merged or not.

$$S_{PCA}^{a,b} = \frac{1}{Q} \sum_{a=1}^{Q} \sum_{b=1}^{Q} \cos^2 \theta_{a,b} = \frac{1}{Q} tr \left(U_{a.q}^T U_{b,Q} U_{b.Q}^T U_{a,Q} \right) \tag{4.19}$$

where $U_{a,Q}$, $U_{b,Q}$ are the first Q eigenvalues of the eigenvectors for segments Sub_S^a and Sub_S^b, respectively.

The distance between the cluster centers of Sub_S^a and Sub_S^b is another merging criterion.

$$D\left(v_o^x, v_p^x \right) = v_o^x - v_p^x \tag{4.20}$$

Since the clustering process is conducted in the whole range of S, the clustering similarity of each state space is measured by the fuzzy decision algorithm proposed by U. Kaymak et al. [40], in which the overall similarity matrix is $\mathrm{H} = \left\{ \mathrm{h_{a,b}} | 1 \le \mathrm{a} \le \mathrm{P}, 1 \le \mathrm{b} \le \mathrm{P} \right\}$,

$$h_{a,b} = \sqrt{\frac{\left(\frac{1}{P(P-1)}\sum_{a=1}^{P}\sum_{\substack{b=1\\ b\neq 0}}^{P} S_{PCA}^{a,b}\right)^2 + \left(\frac{1}{P(P-1)}\sum_{a=1}^{P}\sum_{\substack{b=1\\ b\neq 0}}^{P} D\left(v_a^d, v_b^d\right)\right)^2}{2}} \tag{4.21}$$

If the clustering similarity factor $h_{a,a+1}$ of two adjacent segments Sub_S^a, Sub_S^{a+1} is exceeded the set threshold, Sub_S^a, Sub_S^{a+1} will be merged. It ends when the algorithm termination condition ε is reached.

4.2.2.2 The Time-Varying Markov Model

The automatic segmentation method provides a scientific basis for the state number selection of the Markov process. Moreover, we argue that the transition probability between states is related to the state that the device stays, as well as related to the sojourn time that the equipment stays in the present state. In real-world scenarios, the duration a device remains in a specific state directly correlates with the likelihood of transitioning from that state to another. In other words, the state sojourn time has a discernible impact on the probability of state transitions to a certain degree. Consequently, the state transition matrix (Sojourn time-based state transition probabilities) in relation to the state sojourn time is introduced to establish a time-varying Markov process for real-time reliability evaluation. Furthermore, the proposed Markov process is noted as $M = (\pi, A, ST)$, where π is the initial state probability distribution $\pi = \{\pi_i\}$, $1 \leq i \leq N$, $\pi_i = P[s_1 = i]$, $1 \leq i \leq N$; N is the state number, which is determined by automatic partition method mentioned above; A is the time-varying state transition matrix, $A = \left\{a_{ij}\left(st_t^i\right)\right\}$, $1 \leq i, j \leq N$, $1 \leq st_t^i \leq ST^i$, $a_{ij}\left(st_t^i\right)$ represents the state transition probability of the device transitioning from state i to state j when the device stays in state i at time t, $\sum_{j=1}^{N} a_{ij}\left(st_t^i\right) = 1$, $1 \leq st_t^i \leq ST^i$, $1 \leq i, j \leq N$; ST^i indicates the maximum sojourn time in state i; ST represents the state sojourn time distribution, which is assumed to obey a Gaussian distribution.

Firstly, in accordance with the sample set, the sojourn time distribution of each state is determined, and then the model parameters are estimated by the enhanced Baum–Welch algorithm.

(1) When the model parameters $M = (\pi, A, ST)$ are known, the probability that the device stays in state i for a sojourn time T^* at time t is $\alpha_t\left(i, T_t^{i*}\right) = P\left(q_t = i, st_t^i = T_t^{i*} | M\right)$,

When $t = 1$, the initial forward probability is

$$\alpha_1\left(1, st_1^i\right) = 1 \tag{4.22}$$

$$\alpha_1(j, 1) = \alpha_1\left(1, st_1^i\right)a_{1j}\left(st_1^i\right) \tag{4.23}$$

$$\alpha_1\left(N, st_1^N\right) = \sum_{i=2}^{N} \alpha_1\left(i, st_1^i = 1\right)P\left(st_1^i = 1\right)a_{iN}\left(st_1^i = 1\right) \tag{4.24}$$

When $t = 2, 3, ..., T$, the forward probability recursion formula is

$$\alpha_t(j, 1) = \sum_{i=i}^{N-1} \sum_{st_t^i=1}^{ST^i} \alpha_{t-1}\left(i, st_t^i\right)P\left(st_t^i\right)a_{ij}\left(st_t^i\right) \tag{4.25}$$

$$\alpha_t(j, d) = \alpha_{t-1}\left(j, st_t^i - 1\right) \tag{4.26}$$

$$\alpha_t\left(N, st_t^N\right) = \sum_{i=2}^{N-1} \sum_{st_t^i=1}^{ST^i} \alpha_t\left(i, st_t^i\right)P\left(st_t^i\right)a_{iN}\left(st_t^i\right) \tag{4.27}$$

(2) At time t, when the equipment stays in state i for a sojourn time st_t^i, the backward probability of producing the state sequence in time $T - t$ is $\beta_t\left(i, st_t^i\right) = P\left(s_t = i, st_t^i | M\right)$.

When $t = T$, the initial backward probability formula is

$$\beta_T\left(N, st_T^N\right) = 1 \tag{4.28}$$

$$\beta_T\left(i, st_t^i\right) = P\left(st_t^i\right)a_{iN}\left(st_t^i\right) \tag{4.29}$$

$$\beta_T\left(1, st_1^i\right) = \sum_{j=2}^{N-1} \alpha_{1j}\left(st_1^i\right)\beta_T\left(j, st_1^i\right) \tag{4.30}$$

When $0 < t < T$, the backward probability recursion formula is

$$\beta_t\left(i, st_t^i\right) = P\left(st_t^i\right) \sum_{\substack{j=2 \\ j \neq i}}^{N-1} \alpha_{ij}\left(st_t^i\right)\beta_{t+1}(j, 1) + \beta_{t+1}\left(i, st_t^i + 1\right), 1 < i < N \tag{4.31}$$

$$\beta_t\left(1, st_1^i\right) = \sum_{j=2}^{N-1} a_{1j}\left(st_1^i\right)\beta_t(j, 1) \tag{4.32}$$

On the basis of the forward probability formula and the backward probability formula, the conditional probability formula $S = (s_1, s_2, ..., s_T)$, $s_1, s_2, ..., s_T \in \{i, 1 \leq i \leq N\}$ of the state vector can be obtained as

$$P_S = P(S|M) = \sum_{i=1}^{N} \sum_{st_t^i}^{T_t^{i*}} \alpha_t\left(i, st_t^i\right)\beta_t\left(i, st_t^i\right) \tag{4.33}$$

(3) The re-estimated parameters of time-varying Markov model can be obtained by the formula

$$\begin{aligned} \zeta_t(\mathrm{i}, \mathrm{j}, \mathrm{st}) &= \mathrm{p}\left(\mathrm{s_t} = \mathrm{i}, \mathrm{s_{t'}} = \mathrm{j}, \mathrm{st_t^i} = \mathrm{st}|\mathrm{S}, \mathrm{M}\right) \\ &= \begin{cases} \frac{1}{\mathrm{P_S}}\alpha_t\left(1, \mathrm{st_t^1}\right)\mathrm{a_{1j}}\left(\mathrm{st_t^1}\right)\beta_t(\mathrm{j}, 1), \mathrm{i} = 1 \\ \frac{1}{\mathrm{P_S}}\alpha_t(1, \mathrm{st})\mathrm{p}\left(\mathrm{st_t^i}\right)\mathrm{a_{ij}}\left(\mathrm{st_t^i}\right)\beta_{t+1}(\mathrm{j}, 1), 2 \le \mathrm{i}, \mathrm{j} < \mathrm{N} \\ \frac{1}{\mathrm{P_S}}\alpha_t(\mathrm{i}, \mathrm{st})\mathrm{p}(\mathrm{st})\mathrm{a_{iN}}(\mathrm{st})\beta_{t+1}(\mathrm{j}, \mathrm{st}), 2 \le \mathrm{i} < \mathrm{N}, \mathrm{j} = \mathrm{N} \end{cases} \end{aligned} \tag{4.34}$$

$\zeta_t(i, j, st)$ is the probability with a given model M and state sequence of $S = (s_1, s_2, \ldots, s_T)$, $s_1, s_2, \ldots, s_T \in \{i, 1 \le i \le N\}$, that the performance state of equipment will transit to state j, after a sojourn time st in state i.

$\eta_t(i, st)$ is the probability when the device stays in state i for a sojourn time st at time t,

$$\eta_t(i, st) = \frac{1}{P_S}\alpha_t(i, st)p\left(st_t^i = st\right)\left[\sum_{\substack{j=2 \\ j \ne i}}^{N-1} a_{ij}(st)\beta_{t+1}(j, 1) + a_{iN}(st)\beta_t\left(N, st_t^N\right)\right] \tag{4.35}$$

It can be seen from $a_{ij}(st) = p\left(s_{t+1} = j|s_t = i, st_t^i = st\right)$ that the revaluation formula of $a_{ij}(st)$ is

$$\overline{a}_{ij}(st) = \frac{\sum_{t=1}^{T} \zeta_t(i, j, st)}{\sum_{t=1}^{T} \eta_t(i, st)} \tag{4.36}$$

where $\sum_{t=1}^{T} \zeta_t(i, j, st)$ is noted as the expected transition number from state i to state j, when it stays in state i for a sojourn time st. $\sum_{t=1}^{T} \eta_t(i, st)$ is the sum of expected transition number out of state i, when it stays in state i for a sojourn time st.

Assume that the state sojourn time probability obeys a Gaussian distribution $\left(\mu_i, \sigma_i^2\right)$,

$$\overline{\mu}_\mathrm{i} = \frac{\sum_{\mathrm{t}=1}^{\mathrm{T}} \sum_{\mathrm{st}=1}^{\mathrm{T^{i*}}} \eta_t(\mathrm{i}, \mathrm{st}) \cdot \mathrm{st}}{\sum_{\mathrm{t}=1}^{\mathrm{T}} \sum_{\mathrm{st}=1}^{\mathrm{T^{i*}}} \eta_t(\mathrm{i}, \mathrm{st})} \tag{4.37}$$

$$\overline{\sigma}_{\mathrm{i}}^{2}=\frac{\sum_{\mathrm{t}=1}^{\mathrm{T}}\sum_{\mathrm{st}=1}^{\mathrm{T}^{\mathrm{i}*}}\eta_{\mathrm{t}}(\mathrm{i},\mathrm{st})\cdot\mathrm{st}^{2}}{\sum_{\mathrm{t}=1}^{\mathrm{T}}\sum_{\mathrm{st}=1}^{\mathrm{T}^{\mathrm{i}*}}\eta_{\mathrm{t}}(\mathrm{i},\mathrm{st})}-\overline{\mu}_{\mathrm{i}}^{2} \tag{4.38}$$

(4) Reliability and Remaining Useful Life Prediction

The device in the completely failed segment means that it cannot complete the specified task as required. Consequently, it can be considered that the probability $\zeta_t(i, N, st)$ of the device transferring to the completely failed segment N at time t with a sojourn time st is its unreliability. The sum probabilities for device from the current state transferring to every other states is its reliability.

The reliability of a device at distinct times can be defined as

$$R(t)=1-\zeta_t(i,N,st) \tag{4.39}$$

The failure rate function is noted as $\lambda(t)$, and its life distribution function is as $F(t)$, density function is $f(t)$,$f(t)=F'(t)$, the reliability function is $R(t)$, $F(t)+R(t)=1$, $\lambda(t)=f(t)/R(t)$. For $\forall t$, there is $R(t)>0$. Consequently, it can be considered that $\lambda(t)$ is an estimate of the interval fault conditional probability.

Let L denote the life cycle of the device, and $RUL(t)$ denote the remaining life expectancy of the equipment running normally at time t, from time t to the failure time, there is $RUL(t)=E(L-t|L>t)=(1/R(t))\int_{t}^{\infty}R(x)dx$.

At time t^*, if the device stays in state i with a sojourn time st, its remaining useful life is equal to the sum of the effective remaining sojourn time of running state i and the remaining time in other remaining states (assuming that it traverses $i+1,\ i+2,\ \ldots,\ N-1$ states).

Among them, the effective remaining sojourn time in operating state i is

$$rul_{t^*+st}^{i}=rul^{i}\cdot\left(1-\lambda\left(t^*+st\right)\cdot\Delta t\right)=rul^{i}\cdot\left(1-\frac{\zeta_{t^*+st}(i,j,st)}{\eta_{t^*+st}(i,st)}\right) \tag{4.40}$$

where $rul_{t^*+st}^{i}$ is the effective remaining sojourn time when the device stays in state i with a sojourn time st at time t^*; rul^i is the expected sojourn time in state i, $rul^i=\mu_i$.

Consequently, the remaining useful life for device at time t^*, when it stays in state i with a sojourn time st can be obtained,

$$RUL_{t^*+st}^{i,st}=rul_{t^*+st}^{i}+\sum_{j=i+1}^{N-1}\mu_j \tag{4.41}$$

4.3 Results

Rolling bearing is one of the most widely used general mechanical components in rail vehicles [35]. Besides, only 10–20% of bearings can reach their design life time, in accordance with the statistics. They typically cannot work properly due to fatigue, wear, strain, electrical corrosion, fracture, gluing, etc. Consequently, the bearing is taken as an example to verify the scientifically of the proposed method in this research.

4.3.1 *Experiment Description*

Utilized in this study are the complete life data of bearings obtained from the PRONOSTIA experimental platform, meticulously devised and implemented by the AS2M department within the French FEMTO-ST Institute, as illustrated in Fig. 4.2. The PRONOSTIA experimental platform comprises three principal components: the rotating section, the loading section (applying radial force to the test bearing), and the measuring section.

The operational parameters of the bearing are gauged through the application of radial force, the rotational speed, and the torque, all measured at a sampling frequency of 100 Hz. Additionally, the extent of bearing degradation is assessed using vibration and temperature sensors. The vibration signal is captured by sensors positioned at two locations, forming a 90° angle on both the horizontal and vertical axes. The

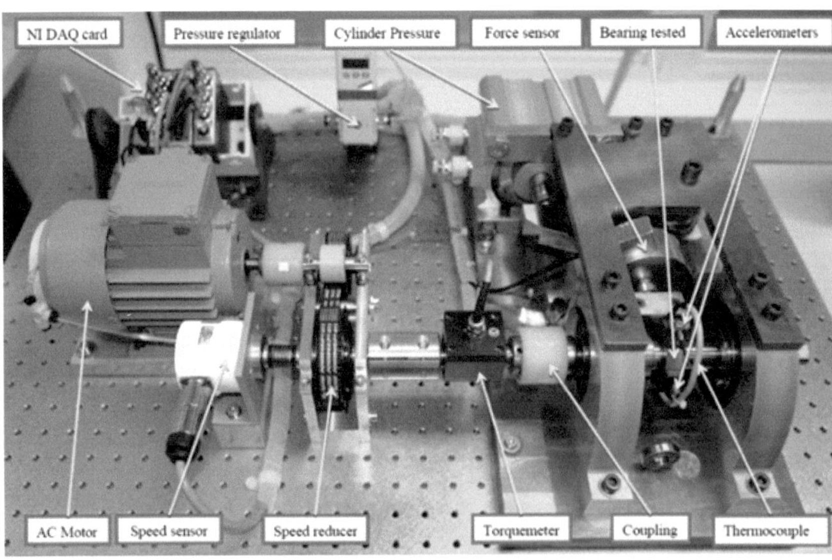

Fig. 4.2 PRONOSTIA test platform of bearing

Table 4.1 Dataset

Dataset	Operating conditions		
	Condition 1	Condition 2	Condition 3
Learning set	Bearing1_1	Bearing2_1	Bearing3_1
	Bearing1_2	Bearing2_4	
	Bearing1_4		
	Bearing1_5		
Test set	Bearing1_6	Bearing2_5	Bearing3_3
	Bearing1_7		

sampling frequency is 25.6 kHz, sampling every 10 s, and each sampling duration of 0.1 s, including 2560 sample points; temperature sensor is located in the hole near the outer ring of the bearing, with a sampling frequency of 10 Hz and 600 points per minute.

The test dataset employed for training in this article is demonstrated in Table 4.1.

4.3.2 Feature Extraction

Calculate the entropy of the bearing vibration acceleration data and align the temperature data with the vibration acceleration data. The Tsallis entropy with $q = 2$, kurtosis, Shannon entropy, and Renyi entropy were extracted from the same full-life of the bearing vibration signal.

As demonstrated in Fig. 4.3, it is obvious that Shannon entropy and Renyi entropy are not monotonic, and they also fluctuate substantially during the full-life signal period. The trend of the kurtosis value appears insufficiently pronounced, and there are instances where values surpass the failure threshold, particularly within the sample range from 450 to 900. These values fail to accurately represent the bearing degradation performance. The Tsallis entropy exhibits monotonic characteristics during the whole life cycle, as well as begins to degrade when the bearing performance is not extremely well (sample number = 1500 in Fig. 4.3), and the Tsallis entropy exhibits good numerical stability in the context of bearing degradation.

The bearing temperature signal is resampled to match the length of the vibration signal. Subsequently, the maximum value is extracted from both signals, and all three groups of signals are individually normalized, converting them into a dimensionless form. Taking Bearing1_1 as an example, its characteristic expression is demonstrated in Fig. 4.4.

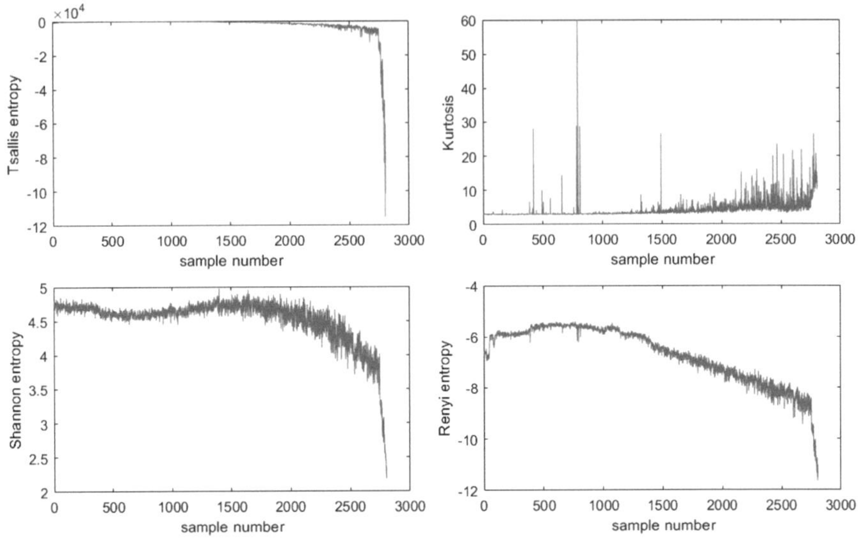

Fig. 4.3 Feature comparison

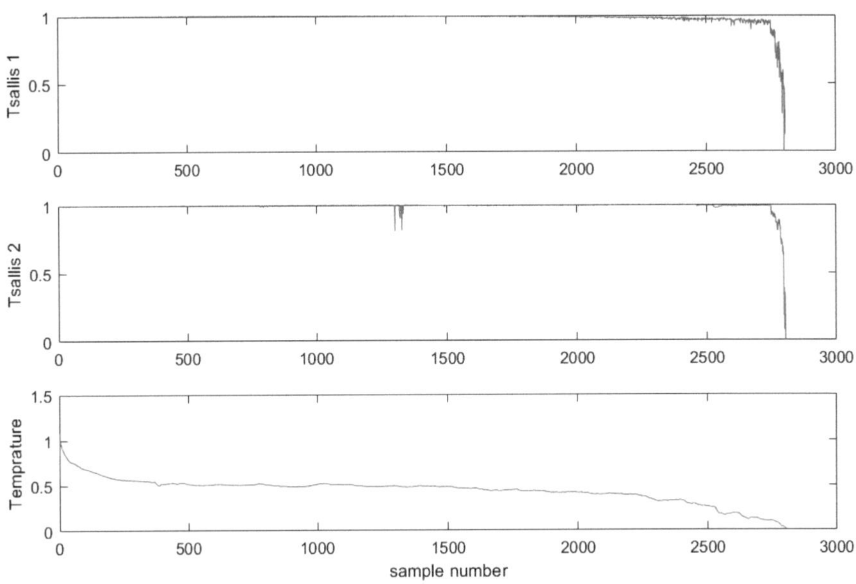

Fig. 4.4 Normalized features of bearing in condition 1

4.3.3 The Automatic State Number Determination of Markov Model

The automatic partition method is performed on the feature data of condition 1-bearing 1 in the dataset. The input feature is [Tsallis 1, Tsallis 2, Temperature], and the result is demonstrated in Fig. 4.5.

Consequently, the segmentation results of Bearing1_1 are [0, 360], [361, 1800], [1801, 2740], [2741, 2803]. In other words, the degradation progression of the bearing can be categorized into four stages, comprising the running-in stage, normal operation stage, degradation failure stage, and complete failure stage. It is consistent with the actual use, which fully verifies the validity and scientific rationality of the multivariate automatic partition algorithm.

The complete failure stage is regarded as the complete failure state; the running-in stage, the normal operation stage and the degenerate failure stage are denoted as the state 1, the state 2, and the state 3 in this paper.

Distinct bearings have distinct degradation trajectories. Samples are aligned on the basis of the slope-based dynamic time warping (DTW) method to acquire more accurate training data, in accordance with the change trend of the degradation amount, not the degradation amount. Illustrated in Figs. 4.6, 4.7 and 4.8 are the optimal path for slope alignment, the results of slope alignment, and the alignment results of samples, respectively, using the alignment process of Bearing1_1 and Bearing1_2 as an example.

The degradation trends of Bearing1_1 and Bearing1_2 are basically the same, as demonstrated in Fig. 4.8. On the basis of the segment location of Bearing1_1, the segment location of Bearing1_2 is ([1,289], [290,392], [393,844], [845,871]).

Nonetheless, not every bearing will go through these four degradation stages. In the tested samples, as demonstrated in Fig. 4.9, the Bearing2_4 fails directly without obvious degradation stage. The break-in stage of Bearing2_4 is defined as [1:402],

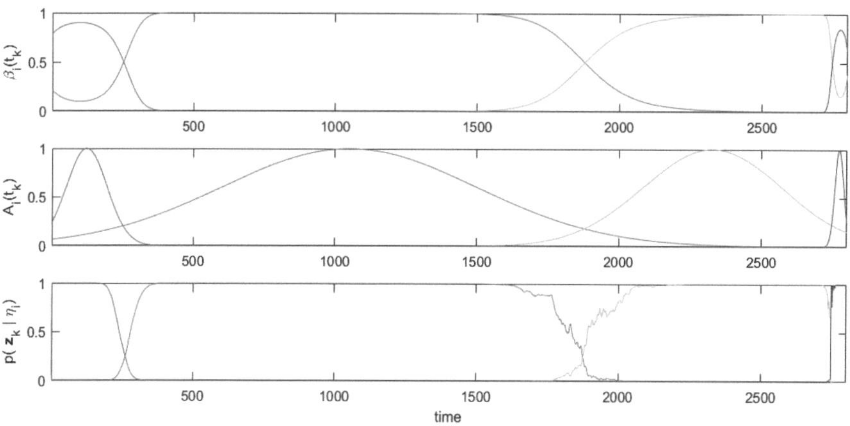

Fig. 4.5 Result of data segmentation

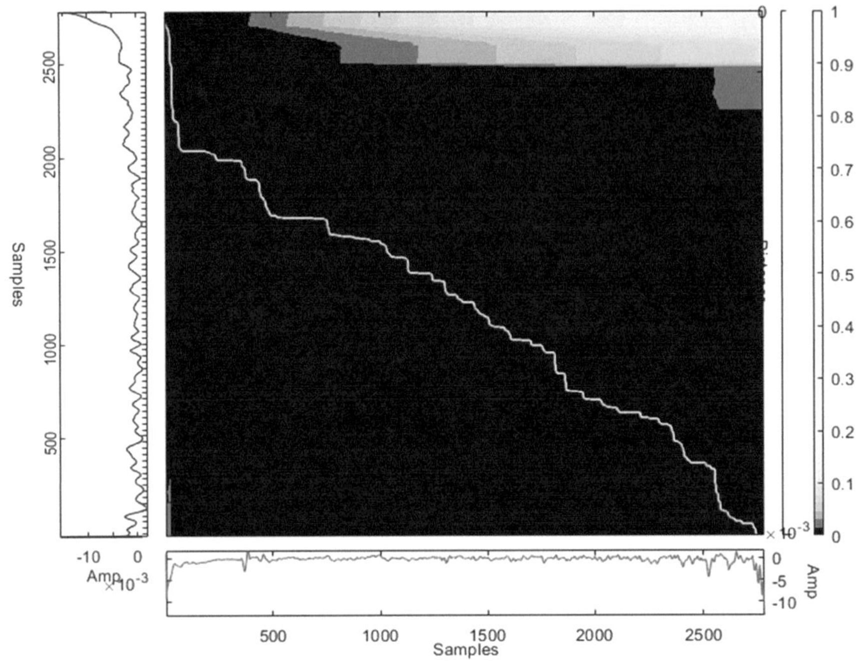

Fig. 4.6 Best path of Slope alignment

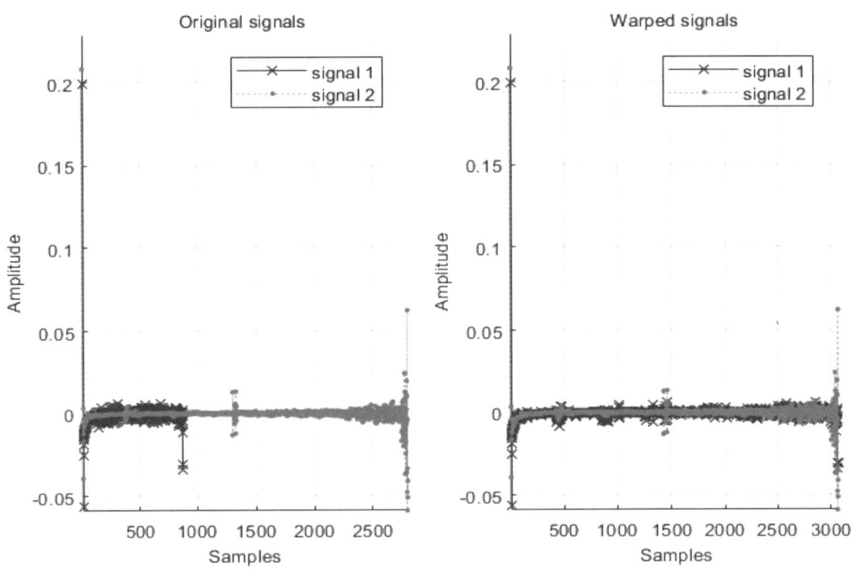

Fig. 4.7 Result of slope alignment

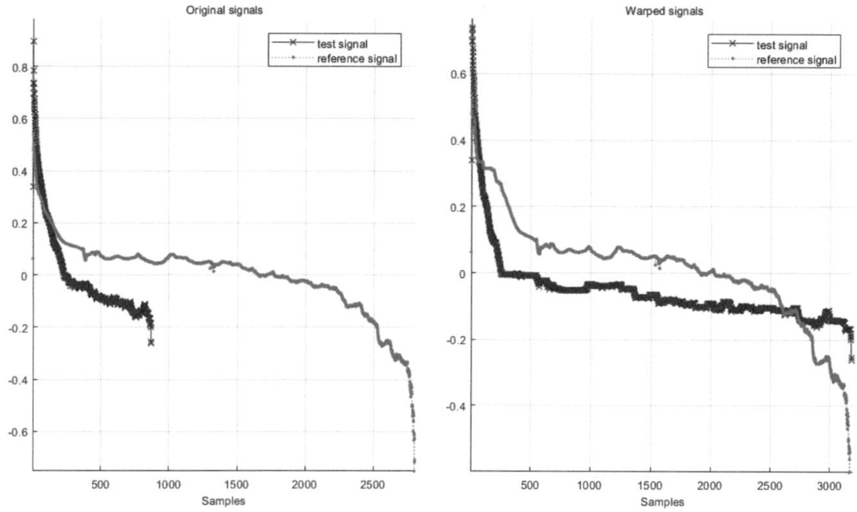

Fig. 4.8 Result of DTW on sample data

the degradation failure stage spans [403, 740], and the complete failure stage extends from [741, 751]. This suggests that the slope-based dynamic time warping method demonstrates excellent performance for samples that do not entirely adhere to the degradation process. It avoids the negative impact arising from confusion of distinct degradation stages during Markov model training. The segmentation result of dataset is listed in Table 4.2.

4.3.4 Reliability Evaluation

To verify the validity and accuracy of the state division of mechanical devices of rail transit trains proposed in this research, Bearing1_1, Bearing1_2, Bearing2_1, Bearing2_4, Bearing3_1 data sets are employed for training, and other bearing data are used for testing.

Within the time-varying Markov process model, it is postulated that the state transition probability adheres to the probability density function of a mixed Gaussian distribution, while the probability distribution function of the state sojourn time follows the probability density function of a single Gaussian distribution. On the basis of the automatic partition method, the state number of the time-varying Markov model is 4, that is $N = 4$.

The maximum number of iteration steps is set to be 1000, and the stop criteria of the error convergence is $1 \times e^{-5}$. The initial matrix of mutual conversion between each state is demonstrated in Table 4.3. Moreover, the average sojourn time in each

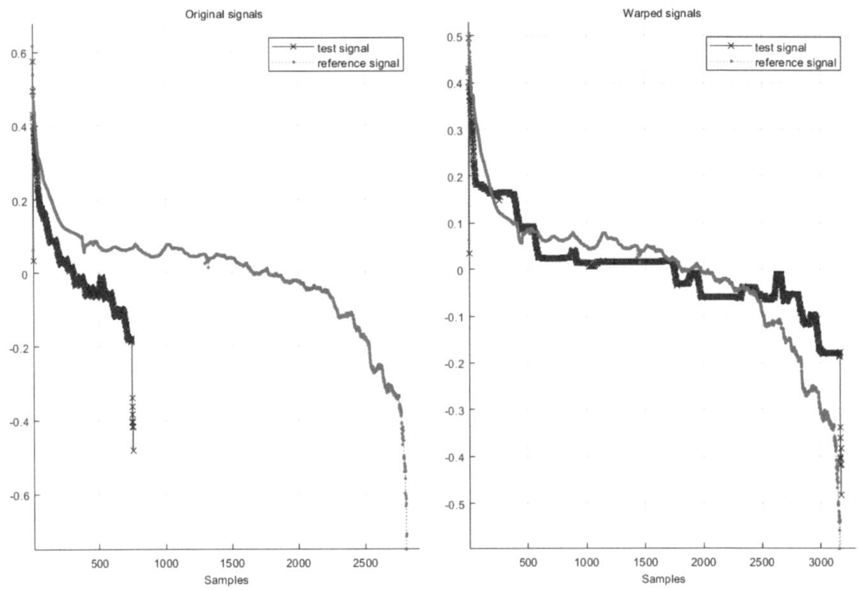

Fig. 4.9 Result of DTW on Bearing2_4 data

Table 4.2 Segmentations result of dataset

Bearing number	State 1 running-in stage	State 2 normal operation stage	State 3 degradation stage	State 4 complete failure stage
Bearing1_1	[1,360]	[361,1800]	[1801,2740]	[2741,2803]
Bearing1_2	[1,289]	[290,392]	[393,844]	[845,871]
Bearing1_4	[1,329]	[330,1084]	[1085,1376]	[1377,1428]
Bearing1_5	[1,612]	[613,1020]	[1021,2421]	[2422,2463]
Bearing1_7	[1,540]	[541,1980]	[1981,2210]	[2211,2259]
Bearing2_1	[1,660]	[661,817]	[818,874]	[875,977]
Bearing2_4	[1,402]	–	[403,740]	[741,751]
Bearing2_5	[1,566]	[567,2254]	–	[2255,2311]
Bearing3_1	[1,246]	[247,490]	–	[491,517]
Bearing3_3	[1,134]	[135,294]	[295,423]	[424,436]

state can be obtained on the basis of training datasets, its result is demonstrated in Table 4.4. A 4-state time-varying Markov process reliability and life prediction model is obtained.

The trained time-varying Markov model is updated on the condition that a new monitoring data sample is arrived. Besides, the bearing staying in state 1 for 10 h is taken as an example. Following feature extraction of the monitoring data sample, the data is sent into the trained time-varying Markov model, and state transition matrix is updated at this time.

Table 4.3 Initial transformation matrix

States	State 1	State 2	State 3	State 4
State 1	0.8712	0.1112	0.0173	0.0003
State 2	0.0000	0.7496	0.2511	0.0993
State 3	0.0000	0.0000	0.7028	0.2972
State 4	0.0000	0.0000	0.0000	1.0000

Table 4.4 Average sojourn time in each state

States	State 1	State 2	State 3	State 4
Average sojourn time	413.8	739.5	383.9	44.3

The bearing reliability at current time is $R(i = 1,\ st_1 = 10) = 1 - 0.0003 = 0.9997$ (Tables 4.5 and 4.6).

To showcase the adaptability of the proposed model to temporal variations, we present the model's changing parameters when the bearing remains in state 2 for a duration of 40 h. At this time, the state transition matrix is demonstrated in Table 4.7, while the mean and variance of the state sojourn time is demonstrated in Table 4.8.

Table 4.5 State transformation matrix when bearing stays in state 1 for 10 h

States	State 1	State 2	State 3	State 4
State 1	0.8695	0.1128	0.0174	0.0003
State 2	0.0000	0.7496	0.2511	0.0993
State 3	0.0000	0.0000	0.7028	0.2972
State 4	0.0000	0.0000	0.0000	1.0000

Table 4.6 The mathematical expectation and variance of sojourn time when bearing stays in state 1 for 10 h

States	State 1	State 2	State 3	State 4
Expectation of sojourn time	390.8451	739.5547	383.93594	44.3543
Variance of sojourn time	2.9458	1.5942	1.5026	1.4358

Table 4.7 State transformation matrix when bearing stays in state 2 for 40 h

State number	1	2	3	4
1	–	–	–	–
2	0.0000	0.6512	0.3291	0.1179
3	0.0000	0.0000	0.7028	0.2972
4	0.0000	0.0000	0.0000	1.0000

Table 4.8 The mathematical expectation and variance of sojourn time when device stays in state 2 for 40 h

States	State 1	State 2	State 3	State 4
Expectation of sojourn time	–	665.9154	383.9326	44.3427
Variance of sojourn time	–	1.9425	1.5026	1.4358

The bearing reliability at current time is $R(i = 2, st_1 = 40) = 1 - 0.1179 = 0.8821$.

For the bearing employed in this experiment, its effective life is the sum of sojourn time in states 1, 2, and 3. When the bearing reaches state 4, it has completely failed. Compared with the hidden Markov model (HMM) algorithm, the prediction result of Bearing1_1 is demonstrated in Fig. 4.10. The lifetime prediction of the HMM presents an obvious step-descent characteristic. Due to the stable state transition matrix of Hidden Markov Model (HMM), the remaining useful life remains unchanged when the device performance is detected in a certain state. This characteristic leads to a significant inconsistency with the actual situation, undermining the model's accuracy.

In contrast, the predicted lifetime of the proposed algorithm in this paper has higher accuracy than HMM, which is in line with the gradual decline with time, since the state transition probability matrix has been designed to update over time theoretically. Additionally, the proposed algorithm becomes more and more accurate as the bearing performance gradually degrades, that is, the results of predicted life are more accurate in the later stage. It should be emphasized that there is also a slight step-descent as the red line indicates in Fig. 4.10, due to the fact that the state transition probability matrix is not changing over time in practice, it is merely updated when a new monitoring data sample is arrived. The shorter the interval of condition monitoring data is, the smoother the prediction curve would be.

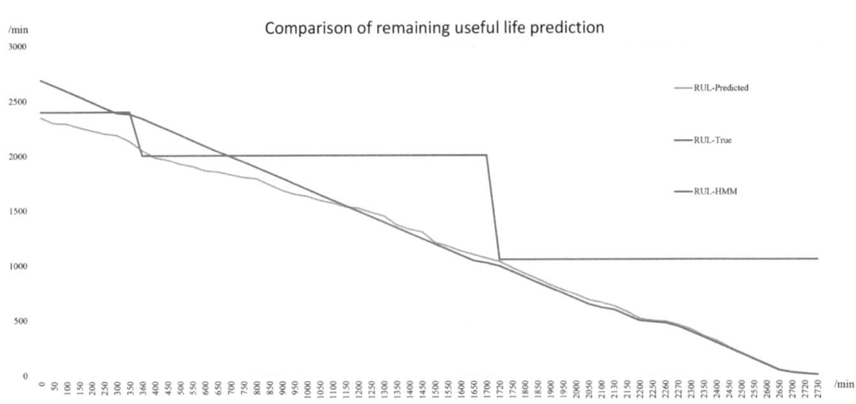

Fig. 4.10 Result of remaining useful life prediction

4.4 Conclusions

The proposed automatic partition time-varying Markov model has two main contributions to real-time reliability evaluation on the basis of the Markov process. Firstly, the state number of the Markov process is selected scientifically on the basis of the device's performance degradation stages. This work is realized automatically, on the basis of the characteristics of the degradation curve, with full use of the multi-source monitoring data. The process is executed autonomously, and the state number is determined by maximizing the utilization of multidimensional data rather than relying solely on a single threshold. This contribution establishes a solid groundwork for the time-varying Markov model. The other main contribution is the time-varying Markov model. It delineates the dynamic progression of device performance degradation in a timely manner. Besides, the polymorphic real-time reliability assessment and remaining useful life prediction are acquired in this procedure. From the experiment, the proposed algorithm demonstrates strong capability for data changes and great potential in accurately performance degradation predicting. Which would be more accurate if the monitoring data points were provided more densely.

There remain some left for enhancement of this work. A good feature is the basis of the whole model. In this work, the feature was manually designed on a vibration signal. Its quality depends entirely on human experience, and this step is also time-consuming. Additionally, the assessment procedure could be simplified in the future. Certain migration algorithms could be viable solutions to address this issue.

Author Contributions Dr. Linlin Kou played a key role in conceptualization, methodology, formal analysis, and writing. Mr. Baiqing Chu served as the Co-supervisor, contributing to the research design, execution, and approval of the final manuscript. Mr. Yan Chen undertook data analysis, result interpretation, and manuscript revisions. Prof. Yong Qin, as another supervisor, contributed to conceptualization, formal analysis, and the final review of the written content. All authors have reviewed and consented to the publication of the manuscript in its final version.

Funding This research was funded by Beijing Mass Transit Railway Operation Corp. LTD., grant number 2021000581000002.

Institutional Review Board Statement Not applicable.

Informed Consent Statement Not applicable.

Data Availability Statement Not applicable.

Conflicts of Interest The authors declare no conflict of interest.

References

1. Arena, S., Roda, I., & Chiacchio, F. (2021). Integrating modelling of maintenance policies within a stochastic hybrid automaton framework of dynamic reliability. *Applied Sciences, 11*, 2300.
2. Bucci, P., Kirschenbaum, J., Mangan, L. A., Aldemir, T., Smith, C., & Wood, T. (2008). Construction of event-tree/fault-tree models from a Markov approach to dynamic system reliability. *Reliability Engineering & System Safety, 93*, 1616–1627.
3. Wang, L., Dai, W., Luo, G., & Zhao, Y. (2019). A novel approach to support failure mode, effects, and criticality analysis based on complex networks. *Entropy, 21*, 1230.
4. Zaman, S. U., Tao, X., Cochrane, C., & Koncar, V. (2021). E-textile systems reliability assessment—A miniaturized accelerometer used to investigate damage during their washing. *Sensors, 21*, 605.
5. Gertsbakh, I. B. (1969). Models of failure. Springer-Verlag.
6. Lvarez, M., Ibáez, J., & Mingo, C. (2020). Reliability assessment of repairable systems using simple regression models. *International Journal of Mathematical, Engineering and Management Sciences, 6*, 180–192.
7. Nejad, R. M., Liu, Z., Ma, W., & Berto, F. (2021). Reliability analysis of fatigue crack growth for rail steel under variable amplitude service loading conditions and wear. *International Journal of Fatigue, 152*, Article 106450.
8. Lehmann, A. (2009). Joint modeling of degradation and failure time data. *Journal of Statistical Planning & Inference, 139*, 1693–1706.
9. Gao, H., Cui, L., & Dong, Q. (2020). Reliability modeling for a two-phase degradation system with a change point based on a Wiener process. *Reliability Engineering & System Safety, 193*, 106601.106601–106601.106609.
10. Tamura, Y., & Yamada, S. (2015). Reliability analysis based on a jump diffusion model with two Wiener processes for cloud computing with big data. *Entropy*.
11. Lin, C. P., Ling, M. H., Cabrera, J., Yang, F., Yu, D., & Tsui, K. L. (2021). Prognostics for lithium-ion batteries using a two-phase gamma degradation process model. *Reliability Engineering and System Safety, 214*.
12. Cai, C. H., Lu, Z. H., Leng, Y., Zhao, Y. G., & Li, C. Q. (2021). Time-dependent structural reliability assessment for nonstationary non-Gaussian performance functions. *Journal of Engineering Mechanics, 147*, 04020145.
13. Meango, J. M., Ouali, & M. S. (2020). Failure interaction model based on extreme shock and Markov processes. *Reliability Engineering and System Safety, 197*.
14. Yeh, W. C. (2021). A quick BAT for evaluating the reliability of binary-state networks. *Reliability Engineering and System Safety, 216*.
15. Rushdi, A., & Ghaleb, F. (2021). Reliability characterization of binary-imaged multi-state Coherent threshold systems. *International Journal of Mathematical, Engineering and Management Sciences, 6*, 309–321.
16. Yeh, W. C. (2021). Computation of the activity-on-node binary-state reliability with uncertainty components.
17. Ding, Y., Hu, Y., Lin, Y., & Zeng, Z. (2021). Reliability analysis of multiperformance multistate system considering performance conversion process. *IEEE Transactions on Reliability*, 1–14.
18. Jiang, S., & Li, Y. F. (2021). Dynamic reliability assessment of multi-cracked structure under fatigue loading via multi-state physics model. *Reliability Engineering & System Safety, 213*.
19. Kuppusamy, S., Joo, Y. H., & Han, S. K. (2021). Asynchronous control for discrete-time hidden Markov jump power systems. *IEEE Transactions on Cybernetics*, 1–6.
20. Chávez-Fuentes, J., Costa, E. F., Terra, M. H., & Rocha, K. (2021). The linear quadratic optimal control problem for discrete-time Markov jump linear singular systems. *Automatica, 127*, Article 109506.
21. Compare, M., Baraldi, P., Bani, I., Zio, E., & Mcdonnell, D. (2020). Industrial equipment reliability estimation: A Bayesian Weibull regression model with covariate selection. *Reliability Engineering and System Safety, 200*.

22. Cao, Y., Liu, S., Fang, Z., & Dong, W. (2020). Modeling ageing effects for multi-state systems with multiple components subject to competing and dependent failure processes. Reliability Engineering and System Safety 2020, 199.
23. He, Y. A., Lc, B., & Nb, C. (2021). New reliability indices for first- and second-order discrete-time aggregated semi-Markov systems with an application to TT&C system. *Reliability Engineering & System Safety*.
24. Zabala, Y. A., & Costa, O. (2020). Static output constrained control for discrete-time hidden Markov jump linear systems. *IEEE Access*, 1–1.
25. Wang, T., Cai, J., Meng, Y., Zhu, S., & Li, Z. (2021). Reliability evaluation method for warm standby embryonic cellular array. *Journal of Ambient Intelligence and Humanized Computing, 12*.
26. Li, M., Kang, J., Sun, L., & Wang, M. (2019). Development of optimal maintenance policies for offshore wind turbine gearboxes based on the non-homogeneous continuous-time Markov process. *Journal of Marine Science and Application, 018*, 1–6.
27. Barbu, V. S., D'Amico, G., & Gkelsinis, T. (2021). Sequential interval reliability for discrete-time homogeneous semi-Markov repairable systems. *Mathematics, 9*.
28. Barbu, V., & Limnios, N. (2008). *Semi-Markov chains and hidden semi-Markov models toward applications*. Springer.
29. Chryssaphinou, O., Limnios, N., & Malefaki, S. (2011). Multi-state reliability systems under discrete time semi-Markovian hypothesis. *IEEE Transactions on Reliability, 60*, 80–87.
30. Moghaddass, R., & Zuo, M. J. (2012). A parameter estimation method for a condition-monitored device under multi-state deterioration. *Reliability Engineering & System Safety, 106*, 94–103.
31. Dong, M., & Peng, Y. (2011). Equipment PHM using non-stationary segmental hidden semi-Markov model. *Robot Cim-Int Manuf, 27*, 581–590.
32. Li, S. (2021). Multi-source knowledge reasoning for data-driven IoT security. *Sensors, 21*.
33. Tsallis, C. (1988). Possible generalization of Boltzmann-Gibbs statistics. *Journal of Statistical Physics, 52*, 479–487.
34. Plastino, A. R., & Plastino, A. (1993). Tsallis' entropy, Ehrenfest theorem and information theory. *Physics Letters A, 177*, 177–179.
35. Himberg, J., Korpiaho, K., Mannila, H., Tikanmaki, J., & Toivonen, H. T. T. (2002). Time series segmentation for context recognition in mobile devices. In *Proceedings of the IEEE International Conference on Data Mining*.
36. Vasko, K. T., & Toivonen, H. T. T. (2002). Estimating the number of segments in time series data using permutation tests. In *Proceedings of the IEEE International Conference on Data Mining*.
37. Bezdek, J. C., & Dunn, J. C. (1975). Optimal fuzzy partitions: A heuristic for estimating the parameters in a mixture of normal distributions. *IEEE Transactions on Computers, C-24*, 835–838.
38. Abonyi, J., Babuska, R., & Szeifert, F. (2002). Modified Gath-Geva fuzzy clustering for identification of Takagi-Sugeno fuzzy models. *IEEE Transactions on Systems Man & Cybernetics Part B Cybernetics A Publication of the IEEE Systems Man & Cybernetics Society, 32*, 612–621.
39. Tipping, M. E., & Bishop, C. M. (1999). Mixtures of probabilistic principal component analyzers. *Neural computation, 11*, 443–482.
40. Kaymak, U., & Babuska, R. (1995). Compatible cluster merging for fuzzy modelling. In *Proceedings of the International Joint Conference of the Fourth IEEE International Conference on Fuzzy Systems*.

Chapter 5
Multistate Reliability Evaluation of Train System on the Basis of Network Flow Theory

5.1 Introduction

Railways become the predominant traffic system carrying a steadily increasing volume of goods and number of passengers in a few decades. This rapid development requires great responsibility for safety and reliability. Meanwhile, Bogie is the main supportive part of rail vehicles and a typical complex mechatronic system that is easily prone to failure. It accounts for a substantial 21.1% on the basis of accumulation of failure data over a couple of years [1]. If a small failure of components occurs, it may result in a loss of production and investments in components or casualties and damage to the system. Consequently, the reliability analysis of the bogie system is an extremely substantial problem and has become one of the most active research fields in railway systems.

Some researchers have made many contributions to the reliability evaluation of rail trains. Numerous traditional methods including fault tree analysis (FTA) [2], Markov analysis [3], Petri net [4], reliability block diagrams (RBD) [5], and failure mode and effect analysis and derivatives (FMEA) [6] were employed to get the reliability of single component on the train.

There is a scarcity of papers on system reliability in this field. Zong et al. addressed this gap by employing node betweenness and cohesion from complex network theory to analyze the significance of specific components and the connectivity of local networks [7]. Wang [8] built a system reliability model on the basis of the connections between components and their life distribution. Connection in distinct types was calculated in distinct ways.

The reliability of the brake and traction system of high-speed rail trains was taken as an example. A copula theory-based reliability model of a dependent component system was investigated and tested on a bogie system in [9]. The relationship of components was taken into consideration and described by copula function. Nevertheless, it is regarded as a binary system. Hu et al. evaluated the reliability of the metro traction power system by fault tree analysis (FTA) and the failure mode and

Y. Qin et al., *Intelligent Safety and Reliability Analysis of Rail Transit Trains*,
https://doi.org/10.1007/978-981-97-4170-0_5

effect analysis (FMEA) [10], and the weakest part of this system was identified through a minimum cut set. Moreover, Lin et al. [11] discussed the reliability of the bogie system on the basis of complex network theory. The topology of the bogie system was investigated with three types of connection between components, which are electrical connection, mechanical connection, as well as information connection. Li et al. [12] proposed a two-state (normal and fault) connectivity reliability of the bogie system with the utilization of a complex network. Nonetheless, these reliability models in the above-mentioned papers were all binary systems, which is an oversimplification of the actual reality [13]. Components and systems are capable of being in a whole range of states, varying from a perfect functioning state, and distinct levels of performance degradation to complete failure. Multistate reliability plays an essential role in our modern society.

Rail train is made of over 40,000 components, with various degradation and complex connection properties. There is an urgent need for an efficient reliability evaluation method. Network models have been applied extensively in numerous real-world multistate systems, including oil/gas production systems [14], computer and communication systems [15], power transmission and distribution systems [16], transportation systems [17], etc. They introduced a more practical approach for addressing multistate system reliability assessment. The capacity-related reliability, defined as the probability of successfully transmitting a specified flow requirement from the input to the output of the network, integrates considerations of path connection, arc capacity, and flow demand [18]. Nonetheless, to the best of our knowledge, it is challenging to determine a network capacity-related reliability model for a mechanical system, particularly a bogie system of high-speed rail trains. The majority of capacity-related reliability evaluations of network models are predominantly utilized in flow-easily-defined systems, including electrical power flow, traffic flow, oil/gas flow, etc.

In this research, we are attempting to build a system reliability evaluation model for the bogie with the consideration of components' multistate through flow network model. The flow of the network model is determined by the component's states. Nodes represent the relationships between components. System reliability is calculated depending on the network d-flow demand, and the rest of this research is organized as follows. Moreover, Sect. 5.2 introduces the multistate network model, and the brief introduction of bogie. Subsequently, Sect. 5.3 indicates the details of the bogie network capacity-related model procedure and its solution. The importance degree of components has likewise been provided in this section, and the final section provides some concluding remarks.

As there is a growing emphasis on providing more reliable services, many systems have integrated reliability analysis as a crucial component in their planning, design, and operation. In this section, the multistate network $G = (N, A, \Omega)$ as demonstrated in Fig. 5.1 and Table 5.1 is employed to illustrate the acronym and notations.

The example indicates $m = 4$, $N = \{1, 2, 3, 4\}$, $A = \{a_1, a_2, a_3, a_4, a_5, a_6\}$, $n = 6$. Arc a_1 has $n_1 = 4$ states 0, 1, 2, and 3 with operating probabilities 0.05, 0.1, 0.25, and 0.6, respectively. The smallest and largest states of a_1, respectively, are $l_1 = 0$ and $u_1 = 3$. Besides, the sets of states of arcs are $K_1 = \{0, 1, 2, 3\}$,

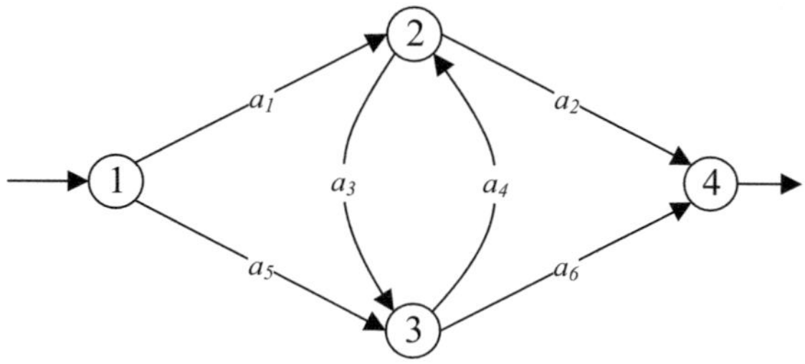

Fig. 5.1 A multistate network [19]

Table 5.1 States and state probabilities of arcs of Fig. 5.1

Arc	States				State probabilities			
a_1	0	1	2	3	0.05	0.10	0.25	0.60
a_2	0	1	2	–	0.10	0.30	0.60	–
a_3	0	1	–	–	0.10	0.90	–	–
a_4	0	1	–	–	0.10	0.90	–	–
a_5	0	1	–	–	0.10	0.90	–	–
a_6	0	1	2	–	0.05	0.25	0.70	–

$K_2 = K_6 = \{0, 1, 2\}$, $K_3 = K_4 = K_5 = \{0, 1\}$. The universal set of state vectors is $\Omega = c = \{(c_1, c_2, c_3, c_4, c_5, c_6)|c_i \in K_i, 1 \leq i \leq 6\}$, and the system state vector $c = (c_1, c_2, c_3, c_4, c_5, c_6) = (3, 2, 1, 0, 0, 1)$ indicates the current states of arcs a_1, a_2, a_3, a_4, a_5, and a_6 are 3, 2, 1, 0, 0, and 1, respectively. Let C be a set of state vectors $C = \{(c_1, c_2, c_3, c_4, c_5, c_6)|1 \leq c_1 \leq 3,\ 1 \leq c_2 \leq 2, 0 \leq c_3 \leq 1, c_4 = c_5 = 0, 0 \leq c_6 \leq 1\}$. Subsequently, $C \subset \Omega$, the smallest state vector in C is $l_C = (1, 1, 0, 0, 0, 0)$, and the largest one is $u_C = (3, 2, 1, 0, 0, 0)$. The sets of states of arcs with respect to C are $K_1^C = \{1, 2, 3\}$, $K_2^C = \{1, 2\}$, $K_3^C = K_6^C = \{0, 1\}$, $K_4^C = K_5^C = \{0\}$, where $K_i^C = \left\{c_i l_i^C \leq c_i \leq u_i^C \text{ for } 1 \leqslant i \leqslant n\right\}$. The maximum flows from source node 1 to sink node 4 under u_C and l_C are $F(u_C) = 3$, $F(l_C) = 1$, respectively.

For a specified demand d, a state vector c is acceptable if $F(c) \geq d$, and c is unacceptable if $F(c) < d$. A set of state vectors C is acceptable (unacceptable) if all state vectors in it are acceptable (unacceptable). On the condition that C is neither acceptable nor unacceptable, it is unspecified. On the condition that the amount of flow sent from s to t is d, the d-flow $= \left(f_1^d, f_2^d, \ldots, f_n^d\right)$ consists of flow f_i^d through arc a_i for each i. Moreover, for state vector $c = (c_1, c_2, \ldots, c_n)$, if $c_i \geq f_i^d \left(c_i < f_i^d\right)$, for $1 \leqslant i \leqslant n$, then c is acceptable (unacceptable). As an example, let $C = \{(c_1, c_2, \ldots, c_6)|1 \leq c_1 \leq 3,\ 1 \leq c_2 \leq 2, 0 \leq c_3 \leq 1, c_4 = c_5 = 0, 0 \leq c_6 \leq 1\}$. For $d = 2$, vector (2, 2, 0, 0, 0, 0) is a 2-flow. The largest state vector $u_c = (3, 2, 1, 0, 0, 1)$ is acceptable due to $F(u_c) = 3 > 2$, while the smallest state vector $l_c = (1, 1, 0, 0, 0, 0)$ is unacceptable due to $F(l_c) = 1 < 2$. The d-flow plays an essential role in dividing C into disjoint subsets. The existence of a d-flow is demonstrated in the following theorem.

Theorem 5.1 For a multistate network $G = (N, A, C)$, if the maximum flow from s to t under the largest state vector in C is no less than d, i.e., $F(u_c) \geq d$, then there exists at least one d-flow.

Due to the fact that flow f_i^d through arc a_i (Step 3 of the proof) is upper bounded by capacity u_i^C (Step 2 of the proof), it is noted here that $f_i^d \leq u_i^c$ for $1 \leq i \leq n$. Nonetheless, it is not guaranteed that $l_i^C \leq f_i^d$ for $1 \leq i \leq n$. In the aforementioned proof, a d-flow can be acquired by examining each arc in A once the maximum flow d from s to t has been established. Over the past three decades, efficient algorithms have been investigated for determining the maximum flow in a flow network [20].

5.2 Bogie Description

Bogie is the driving device for railway vehicles. It basically consists of a bogie frame, wheelset, axle box (with bearing), traction device (consists of motor, shaft coupling, gearbox, etc.), suspension (primary and secondary suspension mechanism), brake equipment, transmission, etc. While there are variations among different bogie types, the main components remain the same as previously mentioned. For a visual representation, please refer to Fig. 5.2 for a direct impression.

The Bogie frame accommodates various bogie equipment, and it is complex in shape and directly supports loads. A wheelset is composed of an axle with bearings and a wheel at each end which is a solid rolled type. The wheelset is susceptible to mass imbalance, which results in vibration of the axle and wheel and is transmitted to the body, producing uncomfortable resonance at certain speeds. The suspension is a crucial subsystem that significantly influences the running performance of the bogie, ride comfort, and the construction of the bogie frame. It plays a key role in absorbing shocks between the axle bearings and the bogie frame, as well as shocks between the bogie frame and the rail vehicle body. This allows for vertical and lateral movement, contributing to the overall performance and comfort of the bogie. Common types are coil springs (primary suspension) and rubber airbags (secondary suspension). The bolster is the main cross member, which is connected to the bogie frame through the secondary suspension and supports the railway car body. Brake equipment is used to stop the train at the desired position.

This research focuses on mechanical braking equipment including wheel tread brakes and disk brakes. In powered vehicles, some form of transmission is typically electrically powered traction motors or a hydraulically powered torque converter [21].

The Bogie serves a number of purposes: support of the rail vehicle body, transferring traction and brake force through adhesion between wheel and rail, stability on both straight and curved track, enhancing ride quality by absorbing vibration and minimizing the impact of centrifugal forces on the condition that the train runs on curves at high speed, as well as minimizing generation of track irregularities and rail abrasion.

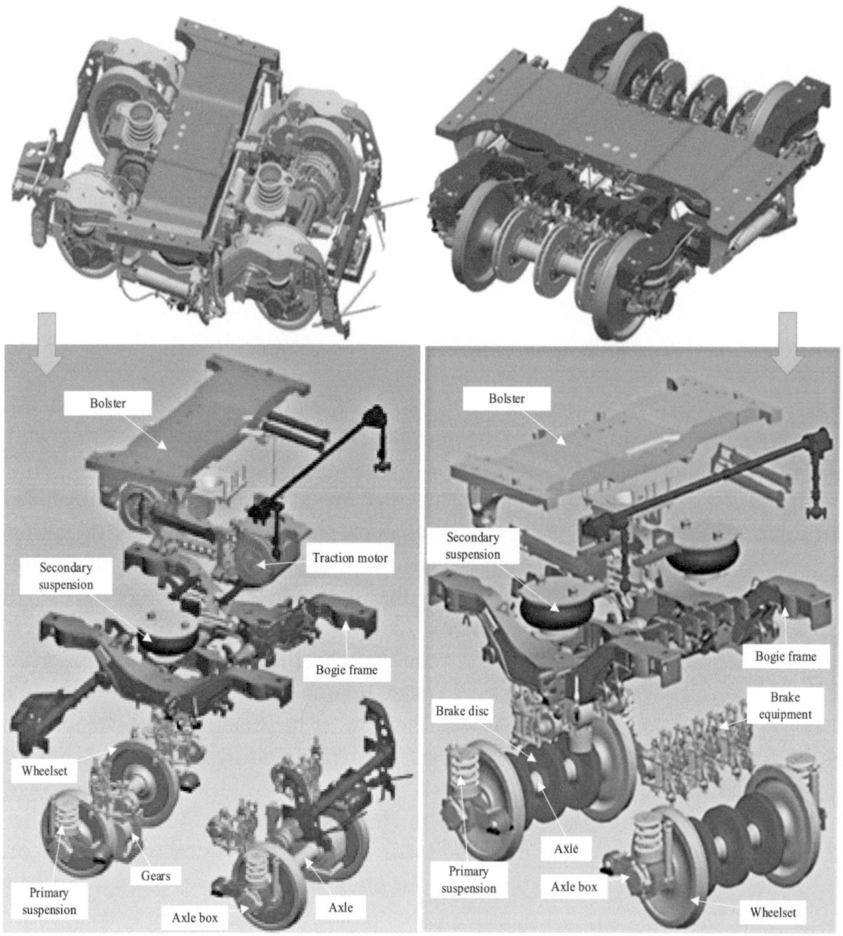

Fig. 5.2 System composition of bogie on CRH3 high-speed rail vehicle

5.3 Modeling

Evaluation of system reliability on the basis of technological networks has been widely used in transport networks [22], road networks [23], electrical networks [24], computer networks [25], social networks [26], etc. These systems are considered as multistate flow networks, which consist of components that can work at distinct levels of capacity. Furthermore, the capacity of each component can take discrete, non-negative integer values following a certain probability distribution. Elements' performances of mechanical systems deteriorate over time, and few components break down from perfect condition immediately. It is a typical property of multi-working conditions on mechanical systems. In this case, the advantages of the flow network model are noticeable.

Nonetheless, to our best knowledge, flow network models for mechanical systems were barely found, especially for bogie systems on high-speed rail trains. Our proposed network flow model which is to deal with multistate reliability evaluation of bogie system with consideration of functional structure is illustrated below.

5.3.1 Network Structure

Essentially, transmission of force, torque, information, electricity, etc. Among distinct components of systems, electromechanical systems, power systems, or any other systems is the energy flow. That makes it possible to normalize distinct substances (force, torque, information, electricity, etc.) into one type, which was generally called flow in this paper. System reliability can be obtained with network flow theory in this means.

The reliability of the bogie on the CRH3 high-speed rail vehicle was evaluated as an example in this section, and only the mechanical parts are taken into consideration. Two problems should be illustrated clearly: What is the flow of the bogie system? What is the path of the flow from the source the node to sink node? Except for supporting the car body mentioned in Sect. 5.2, the main function of the bogie is to transfer traction and braking force between the wheel rail and car body, guarantee safety when railway cars running on a curve, relax impact and attenuate vibration (refer to [27] for details). In our model, the flow of the bogie system was defined as the force that mechanical components transferred and its rate depends on the component's state.

The flow path can be determined by the force track. As the bogie is subjected to a combination of static and dynamic loads, various load conditions are applied. The static load arises from the weight of the body and passenger loads, along with quasi-static loads occurring during curve driving and braking. Additionally, dynamic loads are induced by irregularities in the track, the movement of the rail vehicle body, the driving device, and the motion mode of the wheelset [28]. Generally speaking, there are three types of force that a bogie has to transfer in operation:

(1) The vertical force, which consists of the static load generated by the locomotive weight and the additional dynamic load generated by vertical vibration.
(2) The lateral force, generated by centrifugal force and the horizontal component of train weight due to elevation of outer rail when train on curved track.
(3) The longitudinal force is generated by traction and braking force when the train speeds up or slows down.

The three forces' transmission paths are analyzed to establish a coherent framework for the structure of the bogie network. Some simplification is made to reduce the complexity of the bogie.

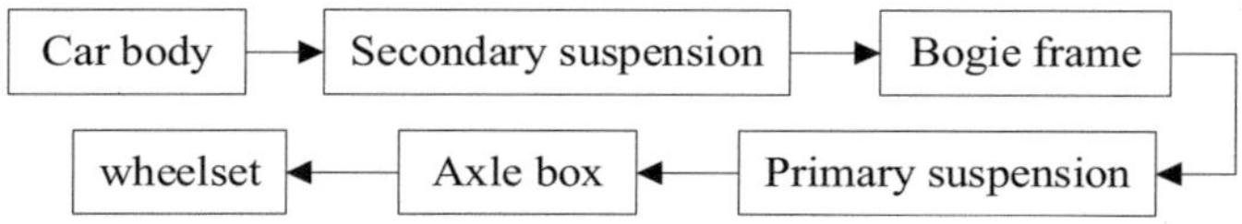

Fig. 5.3 Vertical force transmission paths of bogie

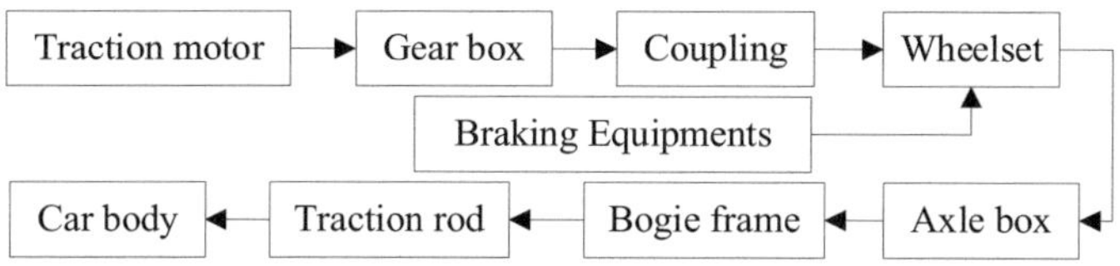

Fig. 5.4 Lateral force transmission paths of bogie

The three force transmission paths are listed below:

(1) The transmission path of the vertical force.
(2) Transmission path of the lateral force.
(3) Transmission path of the longitudinal force.

The structure of bogie network flow model is constructed on the three force routes demonstrated in Figs. 5.3, 5.4 and 5.5. The topology of bogie can be summarized as shown in Fig. 5.6.

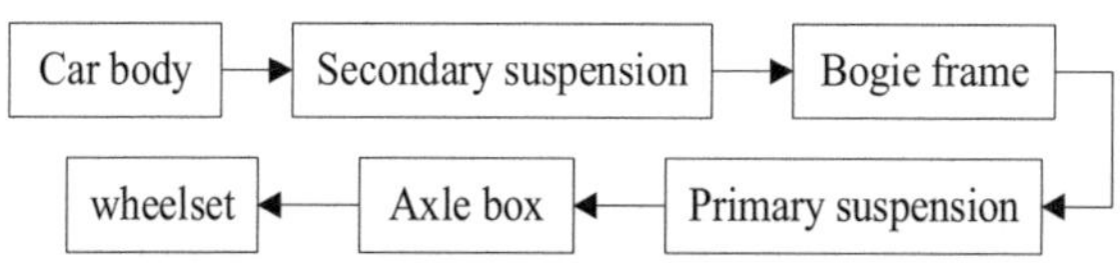

Fig. 5.5 Lateral force transmission paths of bogie

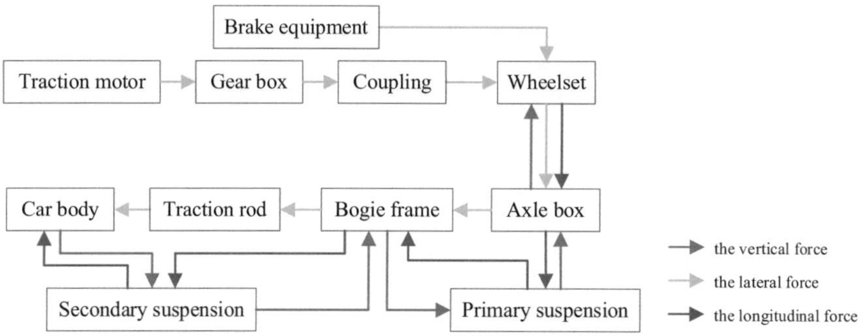

Fig. 5.6 Force transmission paths of bogie

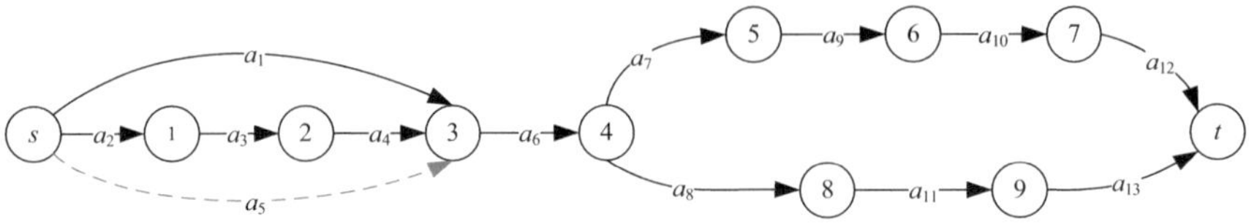

Fig. 5.7 Bogie network model with normalized flow in perfect condition

Due to the fact that the direction of vertical force transmission is opposite to that of the lateral and longitudinal force, we add vertical force into our network structure in the opposite direction according to Newton's Third Law.

Arcs in our network model are set to be the components of a bogie, and the flow on it depends on their degradation level. While the transmission paths between two components are regarded as the nodes. That makes our bogie condition-based reliability model more understandable and computable.

The network structure can be modeled as follows:

where

a_1: braking equipment,

a_2: traction motor,

a_3: gearbox,

a_4: coupling,

a_5: rail,

a_6: wheelsets,

a_7: axle box for longitudinal and vertical force transmission,

a_8: axle box for lateral force transmission,

a_9: primary suspension,

a_{10}: bogie frame for longitudinal and vertical force transmission,

a_{11}: bogie frame for lateral force transmission,

a_{12}: traction rod,

a_{13}: secondary suspension,

s: source node,

t: sink node,

nodes from 1 to 9: the connections between components.

As of now, the foundational bogie network structure has been established, denoted as $G = (N, A, \Omega)$, where $m = 11$, $N = \{s, 1, 2, 3, 4, 5, 6, 7, 8, 9, t\}$, and $A = \{a_1, a_2, a_3, a_4, a_5, a_6, a_7, a_8, a_9, a_{10}, a_{11}, a_{12}, a_{13}\}$. The specifics of each arc's state and the corresponding Ω will be delineated in the subsequent subsection (Fig. 5.7).

5.3.2 Arc Flow Definition

The initial arc flow definition was discussed first, which implies the flow of each arc is primarily defined in the situation that all components and connections between

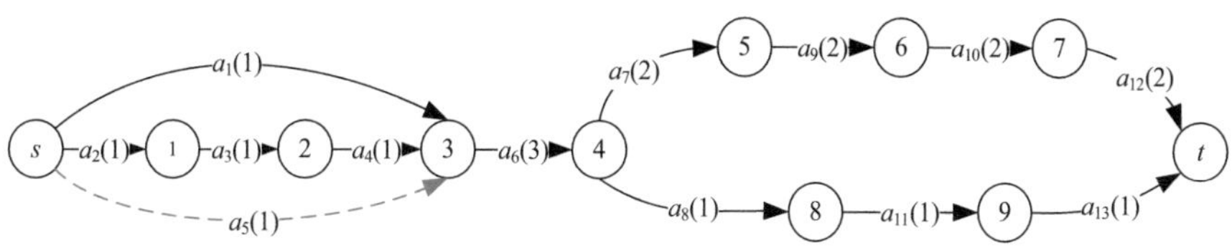

Fig. 5.8 Bogie network model with normalized flow in perfect condition

them are working perfectly. Three kinds of force have to be transferred by the bogie, while some of element work for one or two types of force. The primary flow on each arc is determined by the forces it withstands. A unit flow will be introduced to the arc when it functions for a specific type of force. This methodology has been illustrated in the perfected bogie network model depicted in Fig. 5.8. The blue dotted line symbolizes the rail, responsible for supporting the wheelset in vertical force transmission, and is not considered part of our bogie system. The condition of the rail is consistently assumed to be flawless.

The numbers in parentheses on arcs illustrate the demands that elements need to meet, or the maximum flows components can transfer while working in perfect condition. Nonetheless, performances of all units on the bogie will be dropped after the bogie runs a long time. Invisible degradation exists all the time. The maximum flow of arcs will drop down along with the performance deterioration of all components. On the basis of this idea, the arc flow is determined by the element's state. For instance, if we consider the arc "w" it has the capacity to transfer 3 units of flow under optimal wheelset conditions. In a hypothetical scenario, we can categorize the wheelset degradation into four stages. These stages can accomplish 0, 1, 2, and 3 units of flow, respectively, with associated probabilities of 0.05, 0.1, 0.25, and 0.60.

5.3.3 Multistate Network Reliability Model Building

The general hypothesizes of capacitate d-flow reliability network are as follows:

(1) The network considered is directed and all nodes are perfectly reliable, which means the connections between components of the bogie always work best.
(2) The flows in the network satisfy the so-called flow-conservation law.
(3) There is no demand for the same node, i.e., each node cannot be both the source node and sink node.
(4) The capacity of each component is an independent discrete random variable that may take non-negative integer values following a certain probability distribution (Table 5.2).

The maximum flow capacity of the arc is set to be 10 times larger than its original value in Sect. 5.3.2, as the last assumption asks for an integer value of flow capacity. More choices for the multistate value should be given instead of 0 and 1.

Table 5.2 States and state probabilities of arcs of bogie system

Arc	States				State probabilities			
a_1	0	4	7	10	0.05	0.10	0.25	0.60
a_2	0	5	10		0.10	0.30	0.60	–
a_3	0	3	7	10	0.05	0.10	0.20	0.65
a_4	0	10	–	–	0.10	0.90	–	–
a_5	0	10	–	–	0	1	–	–
a_6	0	18	24	30	0.05	0.25	0.50	0.20
a_7	0	10	16	20	0.10	0.20	0.25	0.45
a_8	0	5	8	10	0.10	0.20	0.25	0.45
a_9	0	12	17	20	0.05	0.05	0.15	0.75
a_{10}	0	16	20	–	0.10	0.20	0.70	–
a_{11}	0	8	10	–	0.10	0.20	0.70	–
a_{12}	0	9	15	20	0.05	0.10	0.15	0.70
a_{13}	0	4	7	10	0.05	0.15	0.35	0.45

There are many algorithms to find out the number of degradation stages together with its probability, including genetic algorithm (GA) [29], support vector machine (SVM) [30], basic random process methods (both Markov and semi-Markov) [31], Mahalanobis–Taguchi system (MTS) [32], multilevel fault-trees [33], etc. The component's multistate reliability evaluation model is beyond our research as this paper is focused on the multistate network reliability model for bogie systems. We make some assumptions about the elements' multistate value and their probabilities. In the light of expression in Sect. 5.2, they are demonstrated as follows.

5.3.4 Solving Procedure of Multistate Network Reliability Model

Theoretically, the evaluation of two-terminal multistate network reliability is an NP-hard problem [34]. Many distinct approaches have been reported to evaluate the exact reliability and its approximation [35–38], and a general one is the two-stage minimal path method [19, 39, 40]. All the minimal path vectors are searched at first, and consequently the exact reliability of two-terminal multistate networks is calculated by using the inclusion–exclusion to evaluate the union probability of these vectors [40]. A recursive algorithm on the basis of breadth-first is utilized to search for all the d-MPs for all possible integer d values [41] together with an enhanced backtracking algorithm proposed by Bai, which is more efficient than other algorithms [42]. Subsequently, an ordering heuristics is employed for reliability evaluation [43].

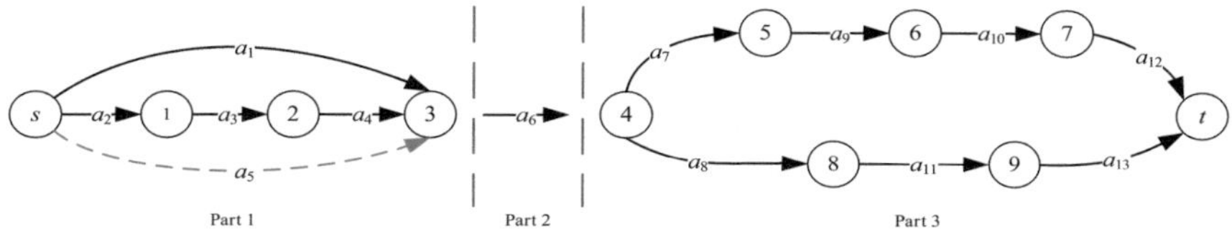

Fig. 5.9 Three parts of bogie network model

We define a simpler one depending on the specific network structure. It is easy to find that a_6 is the bottleneck of the network model, only when a_6 satisfies the d-flow demand, it is possible for the model to meet the requirement. On the basis of this, the whole model is divided into three parts, demonstrated in Fig. 5.9. The only way for the network to suffice is for all three parts to sustain the demand.

The minimal paths (MPs) method [41] and the recursive sum of disjoint products (RSDP) [40] with ordering heuristics [43] are utilized for system reliability calculation. The entire process can be outlined as follows:

Step 1. Divide the network into three parts (as demonstrated in Fig. 5.9).

Step 2. Search for all the d-MPs for all possible integer d values.

We only take Part 1 as an example here to demonstrate the calculation procedure, please refer to [41] for more details.

(1) Input all the vector forms of MPs as follows:

$$\mathrm{MP}_1 = (1, 0, 0, 0, 0);\ \mathrm{MP}_2 = (0, 1, 1, 1, 0);\ \mathrm{MP}_3 = (0, 0, 0, 0, 1).$$

Input maximum state vector $X_{Max} = (10, 10, 10, 10, 10)$. There is no cycle. Set $d = 1$.

(2) Let $d = d + 1$. Build 2-MP candidates by adding MPs to each MP (1-MP) recursively as follows.

For MP_1 with maximum cardinality 1, add MP_1, MP_2, and MP_3 one at a time and obtain the following 2-MP candidates:

$$\overline{X}_{11}^{2} = MP_1 + MP_1 = (2, 0, 0, 0, 0) \tag{5.1}$$

$$\overline{X}_{21}^{2} = MP_1 + MP_2 = (1, 1, 1, 1, 0) \tag{5.2}$$

$$\overline{X}_{31}^{2} = MP_1 + MP_3 = (1, 0, 0, 0, 1) \tag{5.3}$$

$\overline{X}_{11}^{2}$ is the first d-MP candidate built by MPs with the maximum cardinality 1 for level 2;

$\overline{X}_{21}^{2}$ is the first d-MP candidate built by MPs with the maximum cardinality 1 for level 2;

$\overline{X}^2_{31}$ is the first d-MP candidate built by MPs with the maximum cardinality 1 for level 2.

For MP_2 with maximum cardinality 1, add MP_2, and MP_3 one at a time and obtain the following 2-MP candidates:

$$\overline{X}^2_{22} = MP_2 + MP_2 = (0, 2, 2, 2, 0) \quad (5.4)$$

$$\overline{X}^2_{32} = MP_2 + MP_3 = (0, 1, 1, 1, 1) \quad (5.5)$$

$$\overline{X}^2_{33} = MP_3 + MP_3 = (0, 0, 0, 0, 2) \quad (5.6)$$

$\overline{X}^2_{22}$ indicates the second d-MP candidate built by MPs with the maximum cardinality 2 for level 2;

$\overline{X}^2_{32}$ demonstrates the first d-MP candidate built by MPs with the maximum cardinality 2 for level 2;

For MP_3 with maximum cardinality 1, add MP_3 one at a time and obtain the following 2-MP candidates:

$\overline{X}^2_{33}$ illustrates the second d-MP candidate built by MPs with the maximum cardinality 3 for level 2.

(3) Find all feasible 2-MP candidates satisfying all the constraints as follows:

For $\overline{X}^2_{11}$, since $\overline{X}^2_{11} < X_{Max}$, $\overline{X}^2_{11}$ is a 2-MP. Same as $\overline{X}^2_{21}, \overline{X}^2_{31}, \overline{X}^2_{22}, \overline{X}^2_{32}, \overline{X}^2_{33}$.

(4) There are six 2-MP candidates satisfying the constraints, and there are no redundant solutions. Output these 2-MPs as follows:

$$\overline{X}^2_{11} = (2, 0, 0, 0, 0) \quad (5.7)$$

$$\overline{X}^2_{21} = (1, 1, 1, 1, 0) \quad (5.8)$$

$$\overline{X}^2_{31} = (1, 0, 0, 0, 1) \quad (5.9)$$

$$\overline{X}^2_{22} = (0, 2, 2, 2, 0) \quad (5.10)$$

$$\overline{X}^2_{32} = (0, 1, 1, 1, 1) \quad (5.11)$$

$$\overline{X}^2_{33} = (0, 0, 0, 0, 2) \quad (5.12)$$

Go to step 2 (5.2) for next d level.

(5) Let $d = d + 1 = 3$. Build 3-MP candidates by adding MPs to each 2-MP recursively as follows:

For the 2-MP that is built by MPs with maximum cardinality 1, $\overline{X}_{11}^{2}$, add MP_1, MP_2, and MP_3 one at a time and obtain 3 3-MP candidates.

For the 2-MP that is built by the MPs with maximum cardinality 2, $\overline{X}_{21}^{2}$ and $\overline{X}_{22}^{2}$, add MP_2 and MP_3 one at a time and obtain 4 3-MP candidates.

For the 2-MPs that are built merely by the MPs with maximum cardinality 3, $\overline{X}_{31}^{2}$; $\overline{X}_{32}^{2}$ and $\overline{X}_{33}^{2}$, add MP_3 one at a time and obtain 3 3-MP candidates.

(6) Find all feasible 3-MP candidates satisfying all the constraints as follows:

We generate three nonredundant 3-MPs. Display these 3-MPs as follows:

$$(3,0,0,0,0),\ (2,1,1,1,0),\ (2,0,0,0,1),\ (1,2,2,2,0),\ (1,1,1,1,1),$$
$$(1,0,0,0,2),\ (0,3,3,3,0),(0,2,2,2,1),\ (0,1,1,1,2),\ (0,0,0,0,3).$$

Go to step 2. (5.2) for next d level.

Ultimately, we acquire all the ensuing d-MPs.

Step 3. Evaluate the probability of the union of these vector on the basis of the recursive sum of disjoint products (RSDP) [40], and follow the ordering heuristics in reference [43].

The ordering heuristic for d-MPs candidates:

(1) The ordering heuristic O1:

Definition 5.1 Denote Nk as the count of elements shared by a d-MP and the state vector ranked as the kth highest. Subsequently, Nk is defined as the length of this d-MP with respect to the kth highest state.

The ordering heuristic O1 is a hybrid approach that combines cardinality ordering and lexicographic ordering. Here, "cardinality" pertains to the length of d-MP under Definition 5.1, and "lexicographic: refers to the level of the state.

For example, given the 3-MPs in Part 1 of bogie network:

$$z^1 = (3,0,0,0,0),\ z^2 = (2,1,1,1,0),\ z^3 = (2,0,0,0,1),\ z^4 = (1,2,2,2,0),$$
$$z^5 = (1,1,1,1,1),\ z^6 = (1,0,0,0,2),\ z^7 = (0,3,3,3,0),\ z^8 = (0,2,2,2,1),$$
$$z^9 = (0,1,1,1,2),\ z^{10} = (0,0,0,0,3).$$

The highest state vector is defined as

$$ZH^1 = z^1 \oplus z^2 \oplus, \ldots, \oplus z^L \tag{5.13}$$

where $z^1, z^2, \ldots, z^L$ are the L d-MPs which have been given before. $\oplus$ is the same-maximum operator defined in [40].

We have $ZH^1 = (3, 3, 3, 3, 3)$, which implies the highest state is 3 for a_1, a_2, a_3, a_4, a_5. Meanwhile, the second largest state among all 3-MPs for each arc is $ZH^2 = (2, 2, 2, 2, 2)$, the lowest state $ZH^3 = (1, 1, 1, 1, 1)$.

The length of a d-MP with respect to the kth highest state is defined as follows [43], and the length of each 3-MP under Definition 5.1 can be calculated in Table 5.3.

(2) The Ordering Heuristic O2:

Definition 5.2 The length of a-MP is the sum of the values in its logically equivalent vector.

$$z^1 = (3, 0, 0, 0, 0),\ z^2 = (2, 1, 1, 1, 0),\ z^3 = (2, 0, 0, 0, 1),\ z^4 = (1, 2, 2, 2, 0),$$
$$z^5 = (1, 1, 1, 1, 1),\ z^6 = (1, 0, 0, 0, 2),\ z^7 = (0, 3, 3, 3, 0),\ z^8 = (0, 2, 2, 2, 1),$$
$$z^9 = (0, 1, 1, 1, 2),\ z^{10} = (0, 0, 0, 0, 3)$$

The logically equivalent vectors of $z^1, z^2, \ldots, z^{10}$ are

$$\bar{z}^1 = (2, 0, 0, 0, 0), \bar{z}^2 = (1, 1, 1, 1, 0), \bar{z}^3 = (1, 0, 0, 0, 1), \bar{z}^4 = (1, 2, 2, 2, 0),$$
$$\bar{z}^5 = (1, 1, 1, 1, 1), \bar{z}^6 = (1, 0, 0, 0, 2), \bar{z}^7 = (1, 3, 3, 3, 0), \bar{z}^8 = (1, 2, 2, 2, 1),$$
$$\bar{z}^9 = (1, 1, 1, 1, 2), \bar{z}^6 = (1, 0, 0, 0, 3)$$

The length of $z^1, z^2, \ldots, z^{10}$ is 2, 4, 2, 7, 5, 3, 10, 8, 6, 4, respectively.

(3) The Ordering Heuristic O3

In this ordering heuristic O3, in addition to ordering d-MPs on the basis of their lengths, the relationships and similarities among distinct d-MPs are taken into consideration. This approach facilitates the placement of similar d-MPs in close proximity

Table 5.3 Ordering result on the basis of O2

Final order	d-MP	Component state vector					Length by Definition 5.2
1	z^1	3	0	0	0	0	2
2	z^3	2	0	0	0	1	2
3	z^6	1	0	0	0	2	3
4	z^2	2	1	1	1	0	4
5	z^{10}	0	0	0	0	3	4
6	z^5	1	1	1	1	1	5
7	z^9	0	1	1	1	2	6
8	z^4	1	2	2	2	0	7
9	z^8	0	2	2	2	1	8
10	z^7	0	3	3	3	0	10

to each other, with the anticipation of triggering more simplification procedures. Ordering heuristic O4 is directly motivated by the concept of relative complements between two binary MPs and the ordering method reported in [44]. Moreover, by extending the concept of the relative complement in [44] to multistate networks, we define the relative difference for two d-MPs, z^i and z^j, as follows.

The relative distinction between two d-MPs z^i and z^j is

$$\left|\overline{z}^i - \overline{z}^j\right| = \sum_{k=1}^{n} \left\{\overline{x}_i^k - \overline{x}_j^k \middle| \overline{x}_i^k > \overline{x}_j^k\right\} \quad (5.14)$$

where $\overline{z}^i$ and $\overline{z}^j$ are logically equivalent vectors of z^i and z^j, respectively.

L = 10 in our example and d-MP$_{10}$ is z^7. d-MP$_9$ is chosen so that $\left|\overline{z}^{L-1} - \overline{z}^L\right| / \left|\overline{z}^L\right|$ is the largest, where z^L is z^7. The value obtained is given in Table 5.3, and z^8 is chosen as the vector d-MP$_9$ before z^7.

Given the d-MP$_{10}$ and d-MP$_9$, d-MP$_8$ is chosen so that $\sum_{k=0}^{j-1} \left|\overline{z}^{L-j} - \overline{z}^{L-k}\right| / \left|z^{L-k}\right|$ is the largest, where $j = 2$. We obtain the results in Table 5.4, and z^{10} is chosen as d-MP$_8$ to be placed before z^7.

Given the d-MP$_{10}$, d-MP$_9$ and d-MP$_8$, d-MP$_7$ is chosen so that $\sum_{k=0}^{j-1} \left|\overline{z}^{L-j} - \overline{z}^{L-k}\right| / \left|z^{L-k}\right|$ is the largest, where $j = 3$. The outcomes are summarized in Table 5.5, and d-MP$_7$, denoted as z^4, is selected to precede z^{10}.

Given the d-MP$_{10}$, d-MP$_9$, d-MP$_8$ and d-MP$_7$, d-MP$_6$ is chosen so that $\sum_{k=0}^{j-1} \left|\overline{z}^{L-j} - \overline{z}^{L-k}\right| / \left|z^{L-k}\right|$ is the largest, where $j = 4$. We obtain the results in Table 5.5, and z^8 is chosen as d-MP$_6$ to be placed before z^4. The next steps can be conducted as the previous, and they have been eliminated for brevity.

The final order is acquired as $z^6, z^3, z^2, z^5, z^9, z^1, z^8, z^4, z^{10}, z^7$.

Step 4. Evaluate the probability of the union of these vector on the basis of the recursive sum of disjoint products (RSDP) [40], and follow the forehead obtained order in step 3.

Table 5.4 Ordering result after d-MP$_9$ is chosen

Final order	d-MP	Component state vector					Length by Definition 5.2
1	z^1	3	0	0	0	0	3/10
2	z^3	2	0	0	0	1	3/10
3	z^6	1	0	0	0	2	3/10
4	z^2	2	1	1	1	0	2/10
5	z^{10}	0	0	0	0	3	3/10
6	z^5	1	1	1	1	1	2/10
7	z^9	0	1	1	1	2	2/10
8	z^4	1	2	2	2	0	1/10
9	z^8	0	2	2	2	1	1/10
10	z^7	0	3	3	3	0	–

Table 5.5 Ordering result after d-MP$_8$ is chosen

Final order	d-MP	Component state vector					Length by Definition 5.2
1	z^1	3	0	0	0	0	6/10
2	z^3	2	0	0	0	1	5/10
3	z^6	1	0	0	0	2	4/10
4	z^2	2	1	1	1	0	7/10
5	z^{10}	0	0	0	0	3	–
6	z^5	1	1	1	1	1	6/10
7	z^9	0	1	1	1	2	5/10
8	z^4	1	2	2	2	0	8/10
9	z^8	0	2	2	2	1	7/10
10	z^7	0	3	3	3	0	–

The reliability is computed using the subsequent recursive function:

$$\Pr(\emptyset(x) \geq d) \equiv \Pr\left(\left\{x \geq z^1\right\} \cup \left\{x \geq z^2\right\} \cup \ldots \cup \left\{x \geq z^L\right\}\right) \equiv \Pr\left(\bigcup_{i=1}^{L} x \geq z^i\right) \tag{5.15}$$

where z^1, z^2, …, z^L are the L d-MPs which have been given before.

Suppose we have three minimal path vectors z^1, z^2, and z^3. Following the sum of disjoint products (SDP) principle [45]:

$$\Pr \cup \left(z^1, z^2, \ldots, z^L\right) \equiv \sum_{i=1}^{L} TM_i \tag{5.16}$$

where TM_i is the ith term in the SDP calculation [46] given by

$$TM_1 = \Pr\left(\mathrm{x} \geq z^1\right) \tag{5.17}$$

$$TM_1 = \Pr\left(\mathrm{x} \geq z^i\right) - \Pr\left(U_{j=1}^{i-1} x \geq Y^{j,i}\right)$$

$$= \Pr\left(\mathrm{x} \geq z^i\right) - \Pr\bigcup\left(Y^{j,i}, \ldots, Y^{i-1,j}\right), i \geq 2 \tag{5.18}$$

where vector $Y^{j,i} = y^j \oplus y^i$. The length of $Y^{j,i}$ is the same as y^j and y^i, and the value of an element of $Y^{j,i}$ refers to the corresponding state of the corresponding component. The $\Pr \cup (\cdot)$ function with L input vectors can be calculated via $\Pr \cup (\cdot)$ functions with L - 1 input vectors or fewer. The boundary condition is L = 1. In this case, $\Pr \cup (\cdot) = TM_1 = \Pr\left(x \geq y^1\right)$.

$$\Pr(\Phi(x) \geq 3) = \Pr\left(\left\{x \geq Z^1\right\} \cup \left\{x \geq Z^2\right\} \cup \ldots \cup \left\{x \geq Z^{10}\right\}\right)$$
$$= TM_1 + TM_2 + \ldots + TM_{10} \tag{5.19}$$

On the basis of the order, we obtained in step 3:

$$TM_1 = \Pr\left(\text{x} \geq z^6\right) = 0.95 \times 1 = 0.95 \tag{5.20}$$

$$TM_2 = \Pr\left(\text{x} \geq z^3\right) - \Pr\left(\text{x} \geq Y^{6,3}\right) \tag{5.21}$$

where $Y^{6,3} = z^6 \oplus z^3 = (2, 0, 0, 0, 2)$ thus $TM_2 = 0$.

$$TM_3 = \Pr\left(\text{x} \geq z^2\right) - \Pr\left(\left\{\text{x} \geq Y^{6,2}\right\} \cup \left\{\text{x} \geq Y^{3,2}\right\}\right) \tag{5.22}$$

where $Y^{6,2} = z^6 \oplus z^2 = (2, 1, 1, 1, 2)$, $Y^{3,2} = z^3 \oplus z^2 = (2, 1, 1, 1, 1)$.

Since $Y^{6,2} \geq Y^{3,2}$, $Y^{6,2}$ will be removed. Thus, we have

$$TM_3 = \Pr\left(\text{x} \geq z^2\right) - \Pr(\left\{\text{x} \geq Y^{3,2}\right\} = 0 \tag{5.23}$$

$TM_4, TM_5, \ldots, TM_{10}$ can be obtained on the basis of equation (*). Until now, the probability of 3-flow demand of Part 1 has been finished. Subsequent to the flow demand in Part 1, similar procedures can be employed to calculate the flow demand for Part 2 and Part 3.

Step 5. Network probability of *d*-flow demand.

$$Pr^d_{\text{network}} = Pr^d_{\text{Part 1}} \times Pr^d_{\text{Part 2}} \times Pr^d_{\text{Part 3}}\,\text{network}\ 1/4 \tag{5.24}$$

where $\Pr^d_{\text{Part 1}}$, $\Pr^d_{\text{Part 2}}$, $\Pr^d_{\text{Part 3}}$ indicate the reliability of Part 1, Part 2, and Part 3 under the *d*-flow demand, respectively.

The reliability of bogie system with assumption-based capacity distributions is demonstrated in Table 5.6 where R is the reliability of bogie working properly when the *d*-flow demand is required.

Results in Table 5.6 are the probabilities of bogie kept in functional order under a certain condition, and the certain condition is represented as the *d*-flow demand. Take *d*-flow demand $= 13$, $R = 0.6325573$ as an example, if in the hypothetical condition of Table 5.6 (state probability distribution of each component of bogie), with the 13-flow demand, the probability of bogie kept in functional order is 0.6325573. Consider the 13-flow demand as an assumption for the actual operating condition, with parameters set at speed = 300 km/h, loading = 15 t, and 29 °C. The *d*-flow demand of the operating condition will be standardized in future research. Table 5.6 can be utilized for reliability evaluation, even without specific information regarding the current status of the elements.

Additionally, when one or some of the items' status is detected, a more accurate reliability result of the bogie system can be obtained on the basis of our model.

Table 5.6 Bogie system reliability under distinct *d*-flow demands

d-flow demand	1	2	3	4	5	6
R	0.8911012	0.8911012	0.8911012	0.8911012	0.8600548	0.8232590
d-flow demand	7	8	9	10	11	12
R	0.8232590	0.7669154	0.7306946	0.7138445	0.6383528	0.6383528
d-flow demand	13	14	15	16	17	18
R	0.6325573	0.6212220	0.5804271	0.5369206	0.4773950	0.3962036
d-flow demand	19	20	21	22	23	24
R	0.2792655	0.2557102	0.1560475	0.1383053	0.1147509	0.0836373
d-flow demand	25	26	27	28	29	30
R	0.0156992	0.0084373	0.0063942	0.0019747	0.0009873	0.00098737

Table 5.7 shows a further result on the condition that we have already detected wheelset is in perfect status in advance, that is, the probability of $a_6 = 30$ is equal to 1.

There is a slight increase in bogie system reliability from Table 5.6 to 5.7, which can illustrate that our model is suitable for a real system. As when a knowledge of components in the best working condition is given beforehand, a better result should be obtained, intuitively. It also indicates that, even if the elements are in high reliability, the system is not as reliable as we expect it to be, due to the numerous elements and complex system structure.

There are also some inferences that, if the design capacity distributions of components are used for system reliability calculation, the result can be employed as guidance in the design and manufacturing stages, and on the condition that the elements' reliability distribution is obtained from data in operation or prediction, the result is the real-time or forecasting reliability of the bogie system.

Table 5.7 State reliability of bogie system when $a_6 = 30$

State	1	2	3	4	5	6
d-flow	4	5	7	8	9	10
R	0.9380013	0.9053208	0.8665884	0.8072794	0.7691522	0.7514152
State	7	8	9	10	11	12
d-flow	12	13	14	15	16	17
R	0.6719503	0.6658498	0.6539179	0.6109759	0.7670295	0.6819928
State	13	14	15	16	17	18
d-flow	18	19	20	21	22	23
R	0.5660052	0.3989507	0.3653003	0.2229249	0.1975791	0.1639299
State	19	20	21	22	23	24
d-flow	24	25	26	27	28	30
R	0.1194818	0.0784962	0.0421863	0.0319708	0.0098737	0.0049369

Table 5.8 Importance degree of components

Arcs	a_1	a_2	a_3	a_4	a_5	a_6
ID	12.85	7.50	11.80	1.00	0	49.50
Arcs	a_{AB}	a_9	a_{BF}	a_{12}	a_{13}	–
ID	46.50	30.85	28.20	26.85	13.45	–

5.3.5 Importance Degree of Components

It is essential to list components of the bogie system from most important to least important, as the rail train reliability-assignment optimization, operation and maintenance schedules are obtained on the basis of that. The larger the maximum flow, and the lower the reliability of the component, the more important it is. The importance degree of arc ai is defined in Eq. (5.25).

$$ID_{a_i} = \sum c_i \times \left(1 - p_{c_i}\right)\ c_i \in K_i \tag{5.25}$$

The list of elements on the bogie is wheelset > axle box > primary suspension > bogie frame > traction rod > secondary suspension > braking equipment > gear box > traction motor > coupling > rail.

The rationality of this ID equation can be proved in these factors below (Table 5.8):

(1) The importance degree of rail equals 0, as we do not take it into consideration.
(2) The wheelset is the most important element of the bogie system, not only because it undertakes the most flow rate, but also because it is the one with the lowest reliability.
(3) The axle box is the second important component, as it undertakes as much flow rate as the wheelset does, and the reliability is a little bit higher than the wheelset's.
(4) The flow rate distribution of secondary suspension is the same as the braking equipment, but braking equipment is slightly not as reliable as secondary suspension based on their reliability contributions, so secondary suspension is more important than braking equipment.

5.4 Conclusion

This research proposes a multistate reliability evaluation method for rail train bogie, on the basis of a general multistate two-terminal network. It takes bogie functional structure and the components' states into consideration. Arcs at distinct capacity distributions represent component's degradation level, and the d-MP probability of the bogie network system is the reliability of the bogie without any forehead information, which is of great guiding function for reliability design. Furthermore, the system reliability under the given condition can also be provided, together with the importance degree of each component, which can be employed to direct daily system operation and predictive maintenance.

This research provides a more useful, practical, and efficient way to deal with the complex system reliability evaluation, and also clues to extend the usage of network flow theory into describing the working status of mechanical elements.

Subsequent research endeavors will concentrate on exploring the actual capacity distributions of components and establishing a quantitative representation of real working conditions incorporating *d*-flow demand, thereby enhancing the model's practicality.

References

1. Guo, L., Wang, Y. H., & Zhu, L. X. (2013). Failure mode and effect analysis of the bogie for EMU. *Railway Locomotive & Car*.
2. Chen, S. K., Ho, T. K., & Mao, B. H. (2007). Reliability evaluations of railway power supplies by fault-tree analysis. *IET Electric Power Applications, 1*(2), 161–172.
3. Doyon, L. R. (2008). Markov chains in reliability analysis by computer. *Food Additives & Contaminants: Part A: Chemistry, Analysis, Control, Exposure & Risk Assessment. Foreword, 25*(1), 133–140.
4. Lijie, C., et al. (2012). Verification of the safety communication protocol in train control system using colored Petri net. *Reliability engineering & system safety, 100*(Supplement C), 8–18.
5. Hsu, W. C., & Chiu, Y. T. (2012). An application of reliability analysis on the signaling system of the Kaohsiung metro. In *2012 International Symposium on Computer, Consumer and Control*, pp. 564–567.
6. Seo, S. I., et al. (2010). Reliability management and assessment for the electric traction system on the Korea high-speed train. *Proceedings of the Institution of Mechanical Engineers, Part F: Journal of Rail and Rapid Transit, 224*(3), 179–188.
7. Zong, G., Zhang, C., & Wang, H. (2014). Reliability study of the components for the traction system of high speed train based on complex network theory. *China Railway Science, 35*(1), 94–97. (in Chinese).
8. Wang, Y.- H., et al. (2014). Reliability calculation methods based on description of associated equipment system topology relationships and application. *Journal of the China Railway Society, 8*, 32–37. (in Chinese).
9. Lijie, L. I., et al. (2016). Copula theory based calculation of reliability of dependent components system. *China Safety Science Journal, 26*(5), 63–68. (in Chinese).
10. Hu, H.-T., et al. (2012). Reliability evaluation of metro traction power supply system based on FTA and FMEA methods. *Journal of the China Railway Society, 34*(10), 48–54. (in Chinese).
11. Lin, S., et al. (2016). *Reliability study of bogie system of high-speed train based on complex networks theory*, pp. 117–124.
12. Li, L., et al. (2016). Reliability evaluation for complex system based on connectivity reliability of network model. In *2015 International Conference on Logistics, Informatics and Service Sciences (LISS)*.
13. Hudson, J. C., & Kapur, K. C. (1982). Reliability analysis for multistate systems with multistate components. *Microelectronics Reliability, 22*(1), 1–7.
14. Aven, T. (1987). Availability evaluation of oil/gas production and transportation systems. *Reliability Engineering, 18*(1), 35–44.
15. Aggarwal, K. K., Gupta, J. S., & Misra, K. B. (1975). A simple method for reliability evaluation of a communication system. *IEEE Transactions on Communications, 23*(5), 563–566.
16. Ke, W. J., & Wang, S. D. (1997). Reliability evaluation for distributed computing networks with imperfect nodes. *IEEE Transactions on Reliability, 46*(3), 342–349.

17. Unuvar, M., & Prekopa, A. (2009). Bounding the reliability of a stochastic transportation network with random demands and arc capacities. In *EURO Conference.*
18. Ball, M. O., Hagstrom, J. N., & Scott, P. J. (2010). Threshold reliability of networks with small failure sets. *Networks, 25*(3), 101–115.
19. Yeh, W. C. (2005). A novel method for the network reliability in terms of capacitated-minimum paths without knowing minimum-paths in advance. *The Journal of the Operational Research Society, 56*(10), 1235–1240.
20. Ahuja, R. K., et al. (1997). Computational investigations of maximum °ow algorithms. *European Journal of Operational Research, 97*(3), 509–542.
21. Okamoto, I., & Okamoto, I. (1998). How Bogies work. *Japan Railway Transport Review, 18.*
22. Wu, W. W., Ning, A., & Ning, X. X. (2008). Evaluation of the reliability of transport networks based on the stochastic °ow of moving objects. *Reliability Engineering & System Safety, 93*(6), 838–844.
23. Zio, E., et al. (2008). An analytical approach to the safety of road networks. *International Journal of Reliability, Quality and Safety Engineering, 15*(1), 67–76.
24. Lin, Y. K., & Yeh, C. T. (2011). Maximal network reliability for a stochastic power transmission network. *Reliability Engineering & System Safety, 96*(10), 1332–1339.
25. Rai, S., Kumar, A., & Prasad, E. V. (1986). Computing terminal reliability of computer network. *Reliability Engineering, 16*(2), 109–119.
26. Hernandez, H., et al. (2013). Social network analysis via multi-State reliability and conditional; in°uence models. *Reliability Engineering & System Safety, 109*(109), 99–109.
27. Dong, X. M. (2007). How high-speed EMUS works and their structure characters (China Railway Publishing House).
28. Yoon, S. C., et al. (2013). A study on the characteristics of bogie frame materials. *Applied Mechanics and Materials, 420*, 144–148.
29. Xie, M., Dai, Y. S., & Poh, K. L. (2003). Multi-state system reliability (World Scientific).
30. Cui, Y., Shi, J., & Wang, Z. (2015). Multi-state adaptive BIT false alarm reduction under degradation process. *IEEE Transactions on Instrumentation and Measurement, 64*(3), 671–682.
31. Lisnianski, A. (2012). Lz–transform for a discrete-state continuous-time Markov process and its applications to multi-state system reliability.
32. Hu, J., Zhang, L., & Liang, W. (2013). Dynamic degradation observer for bearing fault by MTS–SOM system. *Mechanical Systems and Signal Processing, 36*(2), 385–400.
33. Girard, C., et al. (2016). Emergency response plan: Model-based assessment with multi-state degradation. *Safety Science, 85*, 230–240.
34. Ball, M. O., Colbourn, C. J., & Provan, J. S. (1992). Network reliability. In *Handbooks in Operations Research & Management Science*, pp. 673–762. Springer.
35. Jane, C. C., & Laih, Y. W. (2008). A practical algorithm for computing multi-state two-terminal reliability. *IEEE Transactions on Reliability, 57*(2), 295–302.
36. Jane, C. C., & Laih, Y. W. (2010). Computing multi-state two-terminal reliability through critical arc states that interrupt demand. *IEEE Transactions on Reliability, 59*(2), 338–345.
37. Zhang, T., & Guo, B. (2012) Capacitated stochastic coloured Petrinet-based approach for computing two-terminal reliability of multi-state network. *Journal of Systems Engineering and Electronics, 23*(2), 304–313.
38. Ramirez-Marquez, J. E., & Coit, D. W. (2005). A Monte-Carlo simulation approach for approximating multi-state two-terminal reliability. *Reliability Engineering & System Safety, 87*(2), 253–264.
39. Lin, Y. K., Fiondella, L., & Chang, P. C. (2013). Quantifying the impact of correlated failures on system reliability by a simulation approach. *Reliability Engineering & System Safety, 109*(109), 32–40.
40. Zuo, M. J., Tian, Z., & Huang, H.-Z. (2007). An e±cient method for reliability evaluation of multistate networks given all minimal path vectors. *IIE Transactions, 39*(8), 811–817.
41. Bai, G., Zuo, M. J., & Tian, Z. (2015). Search for all d-MPs for all d levels in multistate twoterminal networks. *Reliability Engineering & System Safety, 142*, 300–309.

42. Bai, G., Tian, Z., & Zuo, M. J. (2016). An improved algorithm for finding all minimal paths in a network. *Reliability Engineering & System Safety, 150*, 1–10.
43. Bai, G., Zuo, M. J., & Tian, Z. (2015). Ordering heuristics for reliability evaluation of multistate networks. *IEEE Transactions on Reliability, 64*(3), 1015–1023.
44. Rai, S., Veeraraghavan, M., & Trivedi, K. S. (1995). A survey of efficient reliability computation using disjoint products approach. *Networks, 25*(3), 147–163.
45. Kuo, W., & Ming, J. Z. (2003). *Optimal Reliability Modeling*. Wiley.
46. Ramirez-Marquez, J. E., Coit, D. W., & Tortorella, M. (2006). A generalized multistate-based path vector approach to multistate two-terminal reliability. *IIE Transactions, 38*(6), 477–488.

Chapter 6
Networked Operation and Maintenance Decision Support Technology for Rail Transit Trains

Trains, being the pivotal equipment in rail transit operations, can significantly compromise the safety of the rail transit system and disrupt normal traffic order and train operation plans when experiencing failures. Consequently, the good performance state of trains is crucial to ensuring the orderly operation of urban rail transit, and timely maintenance and repair of trains is particularly essential. With the increasing demand for passenger transport, the ongoing complexity of train structures, and the constant enhancement of train running speeds, trains need to maintain safety and reliability for a longer time, which poses a severe challenge to the maintenance of rail transit trains.

6.1 Overview

Analysis of maintenance methods employed by rail transit trains in different Chinese cities reveals that, apart from the Hong Kong subway's emphasis on state repair, mainland subways primarily rely on routine maintenance. Regular repairs predominantly refer to the three-level maintenance system of daily inspection, scheduled repair, and major overhaul, which is a planned maintenance strategy based on the bathtub curve theory to determine the length of the maintenance cycle and ensure the safe operation of the train within a certain period. Nonetheless, this planned maintenance strategy has the following two shortcomings: first, due to the inaccurate measurement of the safety status of the train, the regular maintenance cycle is unreasonable, and there may be problems including "insufficient maintenance" or "excessive maintenance," which may lead to the equipment's performance or status after maintenance to be inferior to before, as well as an increase in maintenance costs; second, regular maintenance is applicable only to equipment whose fault pattern aligns with the bathtub curve.

Y. Qin et al., *Intelligent Safety and Reliability Analysis of Rail Transit Trains*,
https://doi.org/10.1007/978-981-97-4170-0_6

For the purpose of avoiding the shortcomings of regular maintenance, some rail transit trains have implemented online monitoring state repair and system repair strategies. Nonetheless, in field research, we have realized that sensors for monitoring the status of train components are typically installed during train manufacturing and are difficult to install on existing trains; the monitoring information of train components is kept confidential by the monitoring system development company, and subway companies typically possess data analysis results rather than raw data, and these circumstances pose challenges in acquiring monitoring information for train components. Consequently, conducting research on online monitoring and state repair of train components becomes a complex task. Additionally, due to the existence of dependencies between train components, the system repair strategy does not consider the dependencies between components, resulting in high system repair costs.

Consequently, the research purpose of this chapter is to formulate a maintenance strategy comprehensively and scientifically for multi-component systems of rail transit trains, explore and determine an optimization method for a train multi-component system state maintenance strategy that considers dependencies, and propose a suitable maintenance strategy optimization model that considers both economic benefits and ensures the reliability, availability, as well as safety of train components.

For simplicity, in this chapter, the term "components" refers to individual elements, which are the smallest maintenance units within a system, each performing a specific function. This includes the running gear subsystem, which forms a unit. On the other hand, "multi-components" represent the entire large system or subsystem, and the term "equipment" serves as the general descriptor for both single component and multi-component systems.

6.1.1 Related Concepts

The specification and examples of reliability, availability, maintainability, and safety for urban rail transit provide a definition of maintainability: "The ability of a given product to complete specified maintenance tasks within a designated time interval under given operating conditions when specified procedures and resources are used for maintenance [1]." In particular, it can be defined as [2] follows:

Definition 6.1 Maintainability of Urban Rail Transit Vehicle Equipment: The ability of urban rail transit vehicle equipment to maintain and restore the specified functional status under specific operating conditions when maintenance is conducted in compliance with the relevant maintenance regulations, using the specified procedures and resources.

This research defines the maintenance cost, maintenance strategy, and maintenance strategy optimization of urban rail transit vehicle equipment as follows:

Definition 6.2 Maintenance Cost of Urban Rail Transit Vehicle Equipment: The cost of maintenance activities including maintenance, repair, and replacement of urban rail transit vehicle equipment.

Definition 6.3 Maintenance Strategy of Urban Rail Transit Vehicle Equipment: A collection of maintenance activity attribute information including maintenance cycle and maintenance method for urban rail transit vehicle equipment.

Definition 6.4 Optimization of Maintenance Strategy for Urban Rail Transit Vehicle Equipment: This pertains to the choice of a specific maintenance strategy under certain conditions and constraints to minimize the maintenance cost of urban rail transit vehicle equipment, ensuring or maximizing its availability and reliability.

6.1.2 Degradation Process Modeling

Currently, the foundation of state maintenance decision-making is the degradation process modeling of components, which can be categorized into two groups: the degradation process modeling on the basis of probability and statistical theory and the degradation process modeling on the basis of stochastic process theory.

(1) The first type of degradation process modeling
The initial type utilizes probability and statistical theory to formulate the relationship function between equipment status and life. It involves calibrating unknown parameters based on measured data and subsequently optimizing maintenance strategies. Typical models include time-delay models, proportional hazard models, and shock models.

    (1) Time-delay models [3] use the concept of delay time to model the relationship between the number of failures in preventive maintenance strategies and maintenance intervals, which is employed in maintenance cost or downtime models to optimize maintenance intervals.
    (2) Proportional hazard models [4] study the relationship model between failure time and covariates.
    (3) Shock models [5] describe maintenance that is typically in relation to equipment working time, service life, as well as monitoring parameters.

(2) The second type of degradation process modeling.
The second type of degradation process modeling is on the basis of stochastic processes. The degradation process of equipment is a stochastic process that is affected by stochastic factors including the environment and stress. This form of degradation process modeling mainly characterizes the process through Levy processes and Markov processes.

(1) Levy processes
   Levy process models [6] are a class of stochastic processes that include Brownian motion, Gamma processes, and Poisson distributions, which have stationary and independent increments and can describe continuous changes in equipment state space.
(2) Markov processes
   Levy processes describe continuous states, while Markov processes [7] can describe both continuous and discrete equipment state spaces.

In comparison with the first type of degradation process modeling on the basis of probability and statistical theory, the second type on the basis of stochastic process theory is comparatively simple in describing the state change process and can solve complex maintenance problems.

Regarding component degradation process modeling, proportional hazard models and Markov processes are commonly employed in trains. For instance, the proportional hazard model is employed to model the reliability and failure rate of each subsystem of the bogie as covariates, and significant covariates affecting the safety of the bogie operation are screened out to obtain the threshold and control limit for the bogie system's maintenance on the basis of cost optimization [8]. Building upon the Markov decision process, this method describes the health degradation process of train traction power equipment over its life cycle. It then optimizes the maintenance cycle and cost for the traction power equipment [9]. A model for the wear of the subway train wheel flange thickness and tread diameter is established by utilizing Markov process theory, and reasonable reprofiling strategies are formulated [10]. In this chapter, we choose the Markov decision process as the modeling theory for maintenance due to the fact that using stochastic processes to describe component degradation processes is more aligned with engineering practice and provides a simpler means to depict the health state of a single component using discrete values.

6.1.3 Interdependence Between Components

The maintenance strategy for multiple components is not simply the sum of individual component maintenance strategies. In practice, there are specific interdependencies among multiple components, which can be categorized into three types: economic interdependence, structural interdependence, and fault interdependence [11].

(1) Economic interdependence

Economic interdependence implies that repairing multiple components together saves costs in comparison to repairing them individually.

(2) Structural interdependence

Structural interdependence means that certain operational components must be replaced or must be dismantled before the faulty component can be replaced or repaired.

(3) Fault interdependence

Fault interdependence refers to the fact that the failure of one component can cause the failure of other components.

Regarding the application of interdependence between components, some scholars have applied it to the maintenance of multiple components of rail transit vehicles. For instance, utilizing fault chains to describe the interdependence between components of a high-speed train, establishing the reliability model of each component on the basis of the fault chain, and reflecting economic interdependence in the maintenance cost model to optimize maintenance time and cost [12]; establishing a model of interdependent failures between components and optimizing the preventive maintenance cycle under the condition of minor repairs [13]. Nonetheless, despite advancements, there remains a gap in research regarding the application of component interdependence in the maintenance of rail transit vehicles. Further investigation is required in this area. Acknowledging the practical context of train maintenance, this chapter comprehensively explores all three types of interdependence between components.

6.1.4 Maintenance Effectiveness

Currently, the majority of maintenance models integrate maintenance effectiveness, typically assuming that equipment will be restored to a condition of "as good as new" or "as good as old" after maintenance. Nonetheless, this assumption is overly idealistic, and in reality, the technical condition of equipment after maintenance is typically somewhere between "as good as new" and "as good as old." In addition, human error or improper maintenance can cause equipment to deteriorate after maintenance. The maintenance effectiveness can be categorized as repair as good as new, repair as non-new, minor repair following the event, maintenance deterioration, and maintenance failure. Besides, the incomplete maintenance modeling methods can be divided into eight types: (p, q) method, (p(t), q(t)) method, enhancement factor method, virtual age method, shock vibration model method, (α, β) method, as well as composite (p, q) method, etc. [14].

Regarding the application of maintenance effectiveness research, the impact of imperfect maintenance on the failure rate of subway train axle boxes can be expressed by introducing the fault rate increasing factor and the service life decreasing factor [15]. This chapter introduces maintenance optimization variables as a means to depict maintenance effectiveness.

6.1.5 *Maintenance Decision Variables, Objectives, and Optimization Methods*

The purpose of maintenance modeling is to obtain optimal objective values under distinct maintenance decision variables through analytical methods or computer simulation methods. Maintenance decision variables encompass factors such as equipment inspection time, maintenance time, and maintenance status threshold. Concurrently, optimization maintenance objectives can involve minimizing equipment failure time, minimizing maintenance cost, maximizing reliability, or maximizing availability. Traditional optimization algorithms include traditional optimization methods including gradient descent and the Lagrangian method, while intelligent optimization algorithms including genetic algorithms, ant colony algorithms, and particle swarm optimization algorithms are employed as essential decision-making methods to achieve optimal objectives. It constitutes a prominent area of focus in the realm of equipment maintenance decision-making research. Through these methods, the static optimal solution of decision objectives can be obtained. Selecting a suitable optimization algorithm on the basis of the characteristics of specific maintenance decision problems can attain superior outcomes with reduced effort.

Regarding maintenance decision variables, objectives, and optimization methods, the maintenance decision variables explored in this chapter encompass the maintenance cycle and maintenance method. The optimization maintenance objective is to minimize the unit time maintenance cost. The maintenance modeling theory adopts the Markov decision process. Consequently, the objective optimization method in this chapter adopts the strategy iteration method.

6.2 Single Component State Maintenance Strategy Optimization Modeling

This study proposes a single component state maintenance optimization method for trains on the basis of fault data. In the first place, the state of the train single component is divided into regions according to the state partitioning method, and subsequently the decision model of single component maintenance cost considering random factors is described by the Markov decision process. Lastly, the decision variables are set as the maintenance method and maintenance cycle. The objective function aims to minimize the long-term average maintenance cost of a single component, with constraints related to reliability, availability, and replacement cost. The primary research areas are as follows:

(1) Introduce the assumptions and optimization outcomes of the model for optimizing the state maintenance strategy of individual components.
(2) Outline the appropriate model for the health state transition process of individual train components.

(3) Analyze the various types of maintenance methods, the scope of maintenance cycles, and the components comprising the maintenance costs for individual train components.
(4) Construct the single component state maintenance strategy optimization model for trains.

6.2.1 Model Assumptions and Optimization Results Format

Assumptions for the model provide the basis for constructing the optimization model, and the presentation of the optimization results serves as a tangible representation of the model's solutions. This section predominantly introduces the following:

1. Model assumptions

   The optimization model for the maintenance strategy of train components is based on the following fundamental assumptions:

   (1) Maintenance activities for a single component will cause downtime, during which the component will not continue to deteriorate, and yet will cause a certain amount of downtime loss.
   (2) The maintenance cost and maintenance time required for minor repairs, status repairs, as well as major repairs increase in turn.
   (3) Minor repairs can restore the component to its pre-failure state, and status repairs can restore the component to a state between the pre-failure state and the brand-new state or to the brand-new state (the degree of repair depends on the maintenance level), as well as major repairs can restore the component to a brand-new state.
   (4) The state transition of a single component is solely contingent on its present state and not influenced by its historical states preceding the current state.
   (5) The deterioration process of a single component is unidirectional. When no one intervenes, the component always transitions from a good state to a worse state.
   (6) Maintenance activities for a single component have the potential to facilitate a transition from a suboptimal state to an improved state. However, there is also an inherent risk that such activities may lead to a transition from a satisfactory state to a less desirable one.
   (7) The current state of a single component can be determined following detection and testing.
   (8) The state of a single component starts from a brand-new state, that is, from the initial state with a reliability of 1 and an availability of 1.
   (9) The maintenance time of a single component is the downtime of the train.

2. Format of optimization results

   Under certain states of train components, maintenance decisions are predominantly determined by the maintenance method and maintenance cycle adopted

Table 6.1 Form of optimized results for maintenance strategies of individual train components

Component status	Maintenance method	Maintenance cycle
1	$R(1)$	$T(1)$
2	$R(2)$	$T(2)$
…	…	…
$N-1$	$R(N-1)$	$T(N-1)$
N	$R(N)$	$T(N)$

in this state. These two determining factors serve as the maintenance decision variables for the optimization model of the train component maintenance strategy under this condition. They collectively form the set of decision variables for the optimization model of the train component maintenance strategy under this state. All possible decision variable sets for this state constitute the train component maintenance strategy set under this state. Furthermore, the maintenance strategy set for train components under all states constitutes the maintenance strategy set for train components.

A maintenance strategy for a train component comprises a set of decision variables for all states of the component. This configuration represents the format of the optimization results for the train component maintenance strategy, as illustrated in Table 6.1. Assuming that there are N states for a train component, the state of the component is i, the maintenance method is $R(i)$, and the maintenance cycle is $T(i)(i = 1, 2, \ldots., N - 1, N)$.

6.2.2 Single Component Health State Transition Probability

Traditional single components only have two binary states: functional (working) and non-functional (faulty). Nonetheless, train single components have over two states and should be examined as multistate components. Besides, the multistate components studied in this research are defined as a finite number of consecutive-state components. Moreover, the essence of elucidating the train component state transition process lies in effectively categorizing the health status of individual components without predefined classification labels and determining the suitable model for capturing the temporal characteristics of the state. This study uses the 1–9 degree method [16] to divide the train single component status, uses the Markov process to describe the single component state transition process, and calculates the single component state transition probability utilizing the analysis results of failure data.

The construction of the train single component health state transition process model predominantly includes three parts: an introduction to the Markov decision process-related theory, a description of the train single component health state space, and the calculation of the train single component health state transition probability.

6.2.2.1 Markov Decision Process

The Markov process [17] is a random process characterized by independence from its prior history, capable of representing the mutual transition dynamics between the various states of a single component. The Markov decision process (MDP) is a sequential decision process on the basis of the Markov process and has dynamic randomness.

Given the stochastic nature of the train single component health state transitions, the Markov process adeptly captures the randomness inherent in the transition dynamics. This aligns well with engineering reality and enables the representation of the health status of a single component through discrete numerical values. Consequently, this study employs the Markov decision process to solve the maintenance decision problem of the train single component state.

MDP can be represented by a five-tuple:

$$\{S, A(i), P(R(i), i, j), C(i, R(i), T(i)), V\} \tag{6.1}$$

The implications of the elements in the quintuple are as follows:

(1) Status set S: The status set of a train component S represents composed of all possible states that the component may appear in, it is denoted as $S\{1, 2, ..., N-1, N\}$. The higher the numerical value of a train component's state, the worse its condition. Specifically:

    When a train component is in state 1, it demonstrates the component is in a brand new state.

    When a train component is in state $N-1$, it implies the component is in a slightly worse state, and yet can still be employed.

    When a train component is in state N, it indicates the component is in a completely failed state, and cannot operate.

(2) Action set $A(i)$: At a certain decision point, when the train component is in state, repair is conducted utilizing repair method $R(i)$ and repair cycle $T(i)$. The set consisting of repair methods and repair cycles under all states is called the action set.

(3) State transition probability $P(R(i), i, j)$: At a certain decision point, when the train component is in state i, and the repair method is $R(i)$, the probability of transitioning to state $\hat{p}_i$ is represented by the state transition probability.

(4) Cost function $C(i, R(i), T(i))$: At a certain decision point, when the train component is in state i, the cost incurred by adopting repair method $R(i)$ and repair cycle $T(i)$ is indicated by the cost function.

(5) Objective function V: The objective function V is also a decision criterion. For any selected maintenance strategy, the goal is to optimize the Markov decision process, which may be expressed in terms of the total maintenance cost or the long-term unit time maintenance cost, among other metrics.

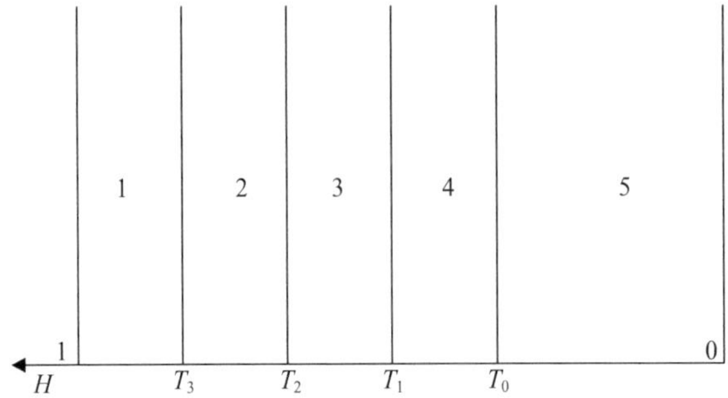

Fig. 6.1 Health status division of individual a single train component

6.2.2.2 Health State Space Partitioning

The widely employed method of multiple-state partitioning is the 1–9 rating method [16], which includes two layers of physical meanings in each level: "safety" and "dangerousness." Odd-numbered levels can ensure the symmetry and complementarity of the two. Three states are overly simplistic, while nine states are excessively intricate. Opting for either five or seven states proves to be more suitable and balanced. Nonetheless, it is challenging to evenly divide the state regions into seven, while five demonstrate effective divisibility, ensuring an even distribution of state regions. Consequently, it is optimal to categorize the health status of train components into five levels.

In this research, the health state score of a single train component H is divided into five regions: $[1, 0.9)$, $[0.9, 0.8)$, $[0.8, 0.7)$, $[0.8, 0.7)$, $[0.7, 0.6)$, $[0.6, 0]$, which are, respectively, illustrated as "normal operation," "mild failure," "moderate failure," "severe failure," and "complete failure," and are, respectively, labeled as $S_1 = 1, S_2 = 2, S_3 = 3, S_4 = 4, S_5 = 5$, as demonstrated in Fig. 6.1. Among them, T_0, T_1, T_2, T_3, respectively, represent the maintenance threshold of the state of the train component. The health state score H can be represented as $H\{[1, 0.9), [0.9, 0.8), [0.8, 0.7), [0.8, 0.7), [0.7, 0.6), [0.6, 0]$.

6.2.2.3 Calculation of the Probability of Train Component Health State Transition

The deterioration process of a train component includes sudden deterioration as well as natural deterioration. Following the failure of a train component, suitable maintenance strategies can be employed to partially or fully restore its functionality. Optimizing the maintenance of a train component can be conceptualized as the inverse process of sudden deterioration, with the degree of optimization (restoration) contingent upon the chosen maintenance level. The process of transition in the health status of a train component under maintenance optimization is demonstrated in Fig. 6.2.

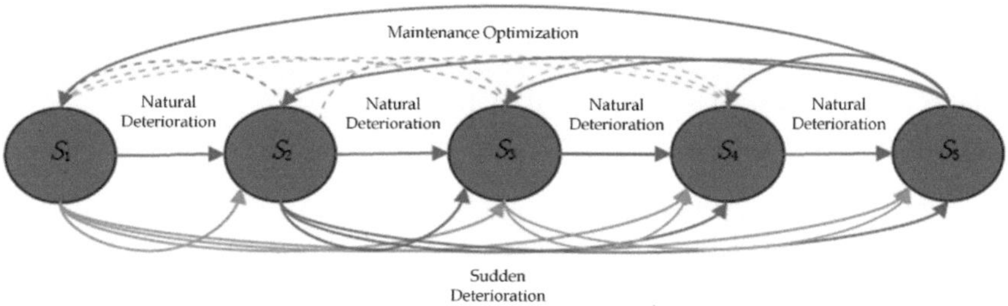

Fig. 6.2 Health state transition process of individual train components considering maintenance optimization

If we assume that both the abrupt deterioration process and the natural deterioration process adhere to a two-parameter Weibull distribution, the degradation process can be expressed in terms of the cumulative degradation amount. The effectiveness of maintenance optimization can then be quantified as the amount of optimized maintenance. Moreover, assuming the initial health score H of the train single component is 1, at the working time t, the total deterioration amount of the single component is $Z(t)$, and the maintenance optimization amount is $Z_R(t)$. Subsequently, the health scores of the single component at the time t can be calculated as follows:

$$H(t) = 1 - \sum_{t=0}^{t} Z(t) + \sum_{t=0}^{t} Z_R(t) \tag{6.2}$$

The model can be articulated through statistical modeling, involving an analysis of the available operational and maintenance data pertaining to train components.

The transition process of the health status of train components can be described by a Markov process, and its transition rule is indicated by the transition probability. According to the C-K equation [18], the calculation formula for the transition probability of a single component's state is as follows [19]:

$$\begin{aligned} P(R(i), i, j) &= \int_{i^-}^{i^+} P\left(j^- \le H(t+\tau) \le j^+ / i^- \le H(t+\tau) \le i^+\right) dH \\ &= \sum_{H \in S} \left(\left(P\left(H(t) \le H^+\right) - P\left(H(t) \le i^-\right)\right)\right. \\ &\quad \times \left(P\left(H(\tau) \le j^+\right) - P\left(H(\tau) \le H^-\right)\right) \end{aligned} \tag{6.3}$$

In the equation, $P(R(i), i, j)$ indicates the probability of a single train component transitioning from a healthy state i to j when repair method $R(i)$ is taken $i, j \in H\{[1, 0.9), [0.9, 0.8), [0.8, 0.7), [0.7, 0.0), [0.6, 0]$.
i^+ represents the upper limit of the health status value i;
i^- denotes the lower limit of the health status value i;
j^+ symbolizes the upper limit of the health status j;
j^- indicates the lower limit of the health status j;
τ implies the degree of discretization of continuous time;
H^+ demonstrates the upper limit of the health state score, which is 0.6;
H^- represents the lower limit of the health state score t, which is 1.

When $j = 5$, $P(R(i), i, 5)$ represents the probability of complete failure of a train component in a specific state i. When $i = 5$ and $j = 5$, it indicates that the train component has attained a state of complete failure, and $P(R(i), i, 5.5) = 1$. When calculating the instantaneous state transition probability, the unit of time is considered as τ. The entirety of state transition probabilities constitutes a matrix known as the state transition probability matrix. The state transition process under maintenance interference is delineated through distinct matrices representing state transitions for different maintenance methods.

Assuming that $Z(t) \sim W(\beta_1, \eta_1)$, the total degradation $Z(t)$ follows a Weibull distribution with (β_1, η_1), and $Z_R(t) \sim W(\beta_2, \eta_2)$, the maintenance optimization quantity $Z_R(t)$, follows a Weibull distribution with (β_2, η_2). When calculating $P\big(H(t) \le H^+\big)$, the formula for $H(t) = 1 - \sum^t Z(t) + \sum^t Z_R(t)i$ substituted in to derive the result. The result of the calculation [20] $P\big(H(t) \le H^+\big)$ is

$$\begin{aligned} P\big(H(t) \le H^+\big) &= P\left(\left(1 - \sum_{t=0}^{t} Z(t) + \sum_{t=0}^{t} Z_R(t)\right) \le H^+\right) \\ &= \sum_{k=0}^{1-H^+} P\Big(\Big(\sum\nolimits_{t=0}^{t} Z(t) + \sum\nolimits_{t=0}^{t} Z_R(t)\Big) \ge 1 - H^+ | Z_R(t) = k\Big) \\ &\quad \times P(Z_R(t) = k) \\ &= 1 - \sum_{k=0}^{1-H^+} \frac{\beta_1 t}{\eta_1 t}\left(\frac{1 - H^+ - k}{\eta_1 t}\right)^{\beta_1 t - 1} \exp\left[-\left(\frac{1 - H^+ - k}{\eta_1 t}\right)^{\beta_1 t}\right] \\ &\quad \times \frac{\beta_2 t}{\eta_2 t}\left(\frac{k}{\eta_2 t}\right)^{\beta_2 t - 1} \exp\left[\left(\frac{k}{\eta_2 t}\right)^{\beta_2 t}\right] \end{aligned} \tag{6.4}$$

6.2.3 Set of Maintenance Methods for Individual Components, Set of Maintenance Periods, and Set of Maintenance Costs

The establishment of the set of maintenance methods, maintenance cycle, and maintenance cost is the foundation for building a maintenance cost model for train components. This section will formulate a range of optional maintenance techniques and maintenance intervals for train components through a comprehensive analysis of literature and on-site research. It will scrutinize the breakdown of maintenance costs and quantify them accordingly.

6.2.3.1 Establishment of Optional Sets for Single Component Maintenance Methods and Maintenance Cycles

The maintenance modes for rail transit trains are typically divided into several types, including daily maintenance, regular inspections, monthly inspections, scheduled maintenance, factory repairs, as well as overhaul repairs. The maintenance interval for each individual train component is established based on either the distance traveled or the operational time. In this research, the maintenance modes for each individual component of the train are divided into four categories: minor repair, Class I status repair, Class II status repair, and major repair.

(1) Minor maintenance: Maintenance performed to maintain normal functions of a single component (including cleaning dirt and adjusting fasteners).
(2) Condition-based maintenance: Targeted maintenance on the basis of the detection and test results of defective components (including non-destructive testing, oil change, dimensional inspection, and parts replacement).
(3) Major maintenance: Replacement of components that meet certain maintenance conditions. Minor maintenance, condition-based maintenance, and major maintenance are categorized under imperfect maintenance, with optimization levels increasing in that sequence.

Consequently, the set of all possible maintenance modes for all individual components of the train is $R = \{\text{minor maintenance, condition-based maintenance, major maintenance}\}$, denoted as $R = \{1, 2, 3, 4\}$.

The set of all possible maintenance cycles for all individual components of the train is $T = \{180\,\text{days},\ 360\,\text{days},\ 540\,\text{days},\ 720\,\text{days},\ 1080\,\text{days}\}$, denoted as $T = \{180, 360, 540, 720, 1080\}$ (excluding daily and monthly inspections).

6.2.3.2 Single Component Maintenance Cost Composition and Quantification

The maintenance cost associated with a single component in a rail transit train encompasses four components: direct maintenance cost, testing and inspection cost, disassembly and assembly cost, and risk cost.

(1) Direct maintenance cost: The cost generated by using maintenance method $R(i)$ when the component is in state i, denoted as $C_M(i, R(i))$, which is in relation to the characteristics of the component, including materials and technology.
(2) Testing and inspection cost: The cost of testing and inspecting the current health status of the single component, denoted as C_T.
(3) Disassembly and assembly cost: Including preparation, logistics, and disassembly and installation costs, which are in relation to the maintenance method and denoted as C_S.
(4) Risk cost: The risk cost generated by adopting maintenance strategy $R(i)$ and maintenance cycle $T(i)$, on the condition that the component is in state i, denoted as $C_{MV}(i, R(i), T(i))$.

The risk cost is predominantly composed of two parts: stochastic failure maintenance cost and downtime loss cost:

(1) Stochastic failure maintenance cost: The unit time maintenance cost spent on sudden deterioration failure maintenance when the component is in state i, denoted as $\rho(i)$.
(2) Downtime loss cost: The downtime loss cost arising from natural deterioration failure maintenance of the component, denoted as C_F.

The aforementioned five fundamental maintenance costs are somewhat abstract and require decomposition to align more effectively with the on-site actualities. The subsequent discussion focuses on the quantification of the aforementioned fundamental maintenance costs [21]:

(1) Quantification of direct maintenance costs, inspection and testing costs, disassembly and assembly costs, as well as random failure maintenance costs.

    The cost elements of the four basic maintenance cost decomposition structures include labor cost and tool/material usage cost:

$$\text{Labor cost} = \text{unit labor cost(CNY/hour)} \times \text{labor time(hours)} \times \text{number of workers.}$$

$$\begin{aligned}\text{Tool and material cost} &= \text{average cost(CNY)} \times \text{number of tools} \\ &\quad + \text{material price(CNY)} \times \text{quantity used.}\end{aligned}$$

(2) The quantification of downtime cost:

$$\text{Downtime cost} = \text{unit downtime cost(CNY/hour)} \times \text{downtime(hours).}$$

$$\begin{aligned}\text{Unit shutdown cost} = {} & \text{passenger capacity} \times \text{occupancy rate} \\ & \times \text{cost per kilometer(CNY/kilometer)} \\ & \times \text{speed(kilometers/hour)}.\end{aligned}$$

6.2.4 *Single Component Maintenance Strategy Optimization Model*

In this research, the maintenance mode and maintenance cycle of train components are taken as decision variables. The optimization model for single component state maintenance strategies for trains is constructed with the minimization of long-term unit time maintenance cost as the primary objective. Reliability, availability, and replacement cost are considered as constraints. The development of the optimization model for single component state maintenance strategies for trains primarily involves the following four processes:

(1) Construction of the train single component maintenance cost model.
(2) Construction of the train single component reliability model.
(3) Construction of the train single component availability model.
(4) Construction of the objective function for optimizing the train single component state maintenance strategy.

6.2.4.1 Single Component Maintenance Cost Model

Prior to formulating the maintenance cost model for an individual train component, it is essential to elucidate the state residence time, random failure maintenance cost, and downtime cost subsequent to the state transition of the said component.

The duration for which a single component of a train remains in a specific state (exceeding the maintenance interval time) can be formulated as follows:

$$T(i) + t_{R(i)} \tag{6.5}$$

where $T_{R(i)}$ represents the maintenance time required for a train component to undergo maintenance method $R(i)$.

When a train component undergoes maintenance method $R(i)$ and maintenance cycle $T(i)$, and transitions from state i to state j, the resulting random failure maintenance cost can be expressed as follows:

$$\sum_{j=1}^{5} P(R(i), i, j) \times \left[\sum_{k=1}^{5} P(1, j, k) \times \left(T(i) + t_{R(i)}\right) \times \rho(i) \right] \tag{6.6}$$

When $P(1, j, k)$ denotes the probability of sudden deterioration failure occurring after the train component transitions to the j state, $j < k$.

When a train component undergoes maintenance method $R(i)$ and maintenance cycle $T(i)$, and transitions from state i to state j, the resulting downtime cost can be expressed as follows:

$$\sum_{j=1}^{5} P(R(i), i, j) \times \left[P(1, j, 5) \times t_{R(i)} \times C_F\right] \tag{6.7}$$

In the equation, $P(1, j, 5)$ indicates the probability of natural deterioration failure occurring following the train component transitions to state $j, j < 5$.

Ultimately, the model for the maintenance cost of a single train component is established by integrating the array of maintenance strategies, maintenance intervals, and fundamental maintenance costs. The expression is

$$\begin{aligned} C_N(i, R(i), T(i)) &= C_M(i, R(i)) + C_T + C_S + C_{MV}(i, R(i), T(i)) \\ &= C_M(i, R(i)) + C_T + C_S + \sum\nolimits_{j=1}^{5} P(R(i), i, j) \\ &\times \left[\sum\nolimits_{k=1}^{5} P(1, j, k) \times \left(T(i) + t_{R(i)}\right) \times \rho(i)\right] \\ &+ \sum\nolimits_{j=1}^{5} P(R(i), i, j) \times \left[P(1, j, 5) \times t_{R(i)} \times C_F\right] \end{aligned} \tag{6.8}$$

In the equation, $C_N(i, R(i), T(i))$ illustrates the total maintenance cost for a specific stage of the equation iteration in step N, when selecting maintenance method $R(i)$ and maintenance cycle $T(i)$ for a single component of the train, and indicating its status as i.

6.2.4.2 Single Component Reliability Model

When the performance level of a train component drops below the user-specified threshold, it is deemed to be in a faulty state, and the performance level of the component becomes a factor in maintenance analysis. Consequently, the reliability of a train component can be measured as a probability [22]:

$$R_N(t) = 1 - P(R(i), i, 5) \tag{6.9}$$

In the equation, $R_N(t)$ represents the reliability of a single component of the train at the iteration step N of the equation.

Additionally, $R_N(t) \geq R_0$, R_0 is the reliability limit value to ensure the normal operation of the train component, which is set at 0.8.

6.2.4.3 Single Component Availability Model

The availability of a train component denotes the ratio of the uninterrupted operating time of the component to the actual operating time and can be articulated as follows:

$$A = 1 - \frac{N_1 \times t_1/3 + N_2 \times t_2 + N_3 \times t_3 + N_4 \times t_4}{\sum T(i) + N_1 \times t_1/3 + N_2 \times t_2 + N_3 \times t_3 + N_4 \times t_4} \tag{6.10}$$

In the equation, $\sum T(i)$ indicates the sum of maintenance periods during the operation time of the train component.
t_1 represents the maintenance time for minor repairs of the train component;
t_2 illustrates the maintenance time for Class I state repairs of the train component;
t_3 demonstrates the maintenance time for Class II state repairs of the train component;
t_4 denotes the maintenance time for major repairs of the train component;
N_1 implies the number of times the train component is subjected to minor repairs before the iteration ends;
N_2 symbolizes the number of times the train component is subjected to Class I state repairs before the iteration ends;
N_3 represents the number of times the train component is subjected to Class II state repairs before the iteration ends; and
N_4 indicates the number of times the train component is subjected to major repairs before the iteration ends.

Except for that, $A \geq A_0$, A_0 is the availability limit value that ensures the normal operation of the train component, which is set to 0.9.

6.2.4.4 Objective Function for Optimizing the Maintenance Strategy of a Single Component

Building upon the aforementioned maintenance cost model for individual train components, a decision-making model for maintenance cost, based on the Markov decision process, is formulated. The Markov decision process can be represented by the Bellman optimality iteration equation [23], and the decision-making model or iteration equation for the maintenance cost of individual train components can be expressed as follows:

$$v_N(i) + \rho_\pi\left(T(i) + t_{R(i)}\right) = C_N(i, R(i), T(i)) + \sum_{j=1}^{5} P(R(i), i, j) \times v_{N-1}{}^*(j) \tag{6.11}$$

In the equation, $v_N(i)$ denotes the relative unit maintenance cost of a single component when the train's single component status is i, at the iteration step N of the equation:

ρ_π indicates long-term unit maintenance cost of a single component of the train;

$v_{N-1}{}^{*}(j)$ symbolizes the minimum relative unit maintenance cost of a single component when the train's single component status is j, at iteration step N of the equation.

Furthermore, the expense incurred for each repair of the train component does not surpass the cost of component replacement. The constraint for the cost of component replacement can be formulated as follows:

$$\rho_{\pi}\left(T(i) + t_{R(i)}\right) \leq v_g \tag{6.12}$$

where v_g represents the cost of component replacement for the train component.

Finally, with the long-term unit time maintenance cost of the train component ρ_{π} as the objective, and the constraints of component reliability, availability, as well as replacement cost, the maintenance cycle and maintenance strategies are considered as decision variables in constructing the optimization model for train component state maintenance strategies:

$$\begin{cases} v_N(i) + \rho_{\pi}\left(T(i) + t_{R(i)}\right) = C_N(i, R(i), T(i)) + \sum_{j=1}^{5} P(R(i), i, j) \times v_{N-1}{}^{*}(j) \\ R_N(t) \geq R_0 \\ A \geq A_0 \\ \rho_{\pi}\left(T(i) + t_{R(i)}\right) \leq v_g \end{cases} \tag{6.13}$$

6.3 Optimization Modeling of Multi-component State Maintenance Strategy

In this research, we propose a state maintenance optimization method for training multi-component systems taking into account the dependence, which can be applied to train multi-component systems with the existence of dependency. This chapter predominantly studies the following seven aspects:

(1) Train multi-component system state maintenance strategy optimization.
(2) Optimization model hypothesis and optimization result form of train multi-component system state maintenance strategy.
(3) Calculation of state transition probability of train multi-component system.
(4) Construction of multi-component maintenance decision model without considering dependency.
(5) Component dependence modeling.

(6) Construction of multi-component maintenance decision model considering dependency.
(7) Construction of objective function of state maintenance strategy optimization for multi-component system.

6.3.1 *Optimization of Multi-component System State Maintenance Strategy*

The construction process of the state maintenance strategy optimization model for a train multi-component system is relatively intricate. Here, the optimization of maintenance strategies for a train multi-component system is segmented into three stages for implementation [24], as demonstrated in Fig. 6.3.

The first stage: The pilot maintenance strategy of the train multi-component system is formulated based on the optimization of the maintenance strategy for individual train components. During this time, the interdependency among components is not taken into consideration, and the maintenance cost of each component is merely aggregated.

The second stage: On the basis of the trial maintenance strategy of train multi-component system, taking into account the impact of economic dependency, structural dependency, and fault dependency on maintenance decisions, the optimization of the maintenance strategy for a multi-component train system with dependencies is undertaken.

The third stage: Finally, update the relevant maintenance information of the train multi-component system to acquire the optimal maintenance strategy of the train multi-component system.

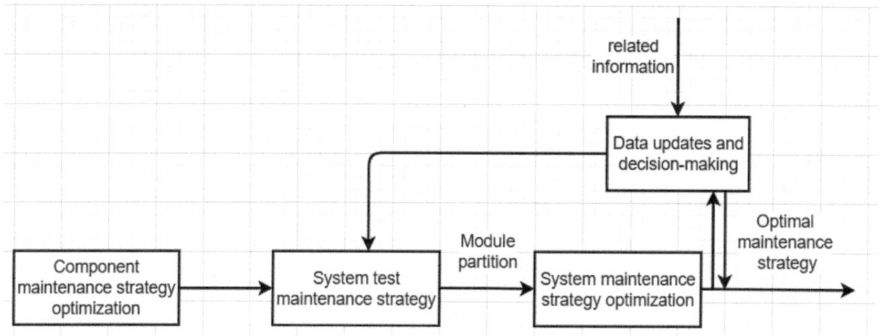

Fig. 6.3 Flowchart of train multi-component system maintenance strategy optimization

6.3.2 *Model Hypothesis and Optimization Results Form*

6.3.2.1 Model Assumptions

In this research, the objects of the optimization model of training multi-component system state maintenance strategy are as follows:

(1) In a multi-component system, all components commence operation from the new state, specifically, the initial state characterized by a reliability of 1 and availability of 1.
(2) At the maintenance time point, all components in the multi-component system are tested simultaneously.
(3) Following multi-component system testing, the current state of all components can be accurately known. For instance, the condition of the axle box bearing can be ascertained by monitoring the shaft temperature, the state of the axle box body can be evaluated based on the axial clearance of the axle box bearing, and the condition of the wheelset can be determined by assessing wheel wear.
(4) In the multi-component system, all components are connected in series in structure.

6.3.2.2 Form of Model Optimization Results

Train multi-component system state is composed of multiple-component states of the multi-component system. In the scenario where the train multi-component system is in a specific state, the maintenance decision is primarily influenced by the maintenance mode and maintenance cycle prescribed for all components within that state. Table 6.2 indicates the form of optimization results of the maintenance strategy of the multi-component train system.

This research assumes that the train multi-component system has M components, and the number of multi-component system states is m, and each component has 5 state, and accordingly the number of multi-component system states is 5^M, that is, $m = 5^M$. When the train multi-component system state is l, the status of part n1 is in $1(l)$, the maintenance mode is $Rn1(l)$, and the maintenance cycle is $T(l)$ $(n_1 = 1, 2, \ldots, M; l = 1, 2, \ldots, m)$.

6.3.3 *State Transition Probability Calculation of Multi-component System*

During the detection time point of the train multi-component system, the current state of all train components can be precisely determined through the multi-component system detection test. The state of all components constitutes the multi-component system state, which can be expressed as a state array $[M_1, M_2, \ldots, M_M]$

Table 6.2 Maintenance strategy optimization results of multi-component train system

System state	Part 1 (state, maintenance mode)	Part 2 (state, maintenance mode)	…	Part M-1 (state, maintenance mode)	Part M (state, maintenance mode)	Maintenance cycle
1	$(i_1(1), R_1(1))$	$(i_2(1), R_2(1))$	…	$(i_{M-1}(1), R_{M-1}(1))$	$(i_M(1), R_M(1))$	$T(1)$
2	$(i_1(2), R_1(2))$	$(i_2(2), R_2(2))$	…	$(i_{M-1}(2), R_{M-1}(2))$	$(i_M(2), R_M(2))$	$T(2)$
…	…	…	…	…	…	…
m−1	$(i_1(m-1), R_1(m-1))$	$(i_2(m-1), R_2(m-1))$	…	$(i_{M-1}(m-1), R_{M-1}(m-1))$	$(i_M(m-1), R_M(\mathrm{m}-1))$	$T(m-1)$
m	$(i_1(m), R_1(m))$	$(i_2(m), R_2(m))$	…	$(i_{M-1}(m), R_{M-1}(m))$	$(i_M(m), R_M(\mathrm{m}))$	$T(m)$

$[M_1, M_2, \ldots, M_M]$ indicate separately the state of part $1, 2, \ldots, M$). cThe state thresholds of each component crossing each other are divided into the multi-component system state space, and divergent areas of the space correspond to distinct state arrays.

Assume that part n_1 is repaired in state $R(i)$. The probability of moving from state i to state j is $dy[n_1, R(i)][i*N + j]$ ($n_1 = 1, 2, \ldots, M$, M shows the number of parts; $i, j = 1, 2, \ldots, N$, N indicates the number of component states), that is, the component state transition probability is stored in the array.. $dy[n_1, R(i)][i*N + j]$ [25].

Store train multi-component system state array $[M_1, M_2, \ldots, M_M]$ in $xt\_zt[n*M + n_1]$ ($l_1 = 1, 2, \ldots, m$, m indicates the number of multipart system state arrays,$m = N^M$; $n_1 = 1, 2, \ldots, M$; M demonstrates the number of parts) [25].

The calculation of the state transition probability in a multi-component system relies on the component state transition probability and the state array of the multi-component system. The multipart system state transition probability array $xt[l*m + n]$ stores the probability of a multipart system moving from state l to state n (l, n = 1, 2,..., m; m indicates the number of multipart system state arrays). The specific steps to calculate $xt[l*m + n]$ are as follows:

(1) Starting with R and i, take an element $xt[l*m + n]$ of the multi-component system state transition probability array.
(2) In the corresponding multipart system state array $xt\_zt[l_1*M + n_1]$, take out rows l and n, i.e., $xt\_zt[l*M + n_1]$ and $xt\_zt[n*M + n_1]$.
(3) When $l \neq n$, the formula for calculating the probability of the multi-component system transitioning from state l to state n is as follows:

$$xt[l*m + n] = \prod_{n_1=1}^{M} dy[n_1, R(i)][[xt\_zt[l*M + n_1] * N + xt\_zt[n*M + n_1]]. \tag{6.14}$$

In formula: $0 < dy[n_1, R(i)][xt\_zt\,[l*M + n_1]*N + xt\_zt\,[n*M + n_1]] < 1$, when the states of two or more components in the state $xt\_zt[l*M + n_1]$ and $xt\_zt[n*M + n_1]$ of the multi-component system are distinct, the probability of the multi-component system transferring from state l to state n is 0, that is, the assumption is made that simultaneous state transfers cannot occur among components.
(4) When $l = n$, the formula for calculating the probability of the multi-component system transferring from state l to state n is as follows:

$$xt[l*m + l] = 1 - \sum_{n=1, m \neq l}^{m} xt[l*m + n] \tag{6.15}$$

(5) $l = 1, 2, \ldots, m$ repeat Step (1)–Step (4) and obtain the state transition probability matrix of the multi-component system:

$$\begin{array}{c} \\ xt\_zt[1*M+n_1] \\ xt\_zt[2*M+n_1] \\ \dots \\ xt\_zt[m*M+n_1] \end{array} \begin{array}{c} xt\_[1*M+n_1]\ xt\_zt[2*M+n_1]\ \dots\ xt\_zt[m*M+n_1] \\ \begin{bmatrix} xt[1*m+1] & xt[1*m+2] & \dots & xt[1*m+m] \\ xt[2*m+1] & xt[2*m+2] & \dots & xt[2*m+m] \\ \dots & \dots & \dots & \dots \\ xt[m*m+1] & xt[m*m+2] & \dots & xt[m*m+m] \end{bmatrix} \end{array} \tag{6.16}$$

On the condition that the state of train multi-component system is l, the state of part n_1 is i; when the state of train multi-component system is n, the state of part n_1 is j; when part n_1 adopts maintenance mode $R(i)$, the probability of transferring from state i to state j is $P^{(n_1)}(R(i), i, j)$. At this time, the state i of component n_1 can also be expressed as $xt\_ztl*M + n_1$, the state j of component n_1 is specified as $xt\_ztn*M + n_1$, the probability $P^{(n_1)}(R(i), i, j)$ of component n_1 transferring from state i to state j when maintenance mode $R(i)$ is adopted can also be expressed as $dy[n_1, R(i)][[xt\_zt[l*M + n_1]*N + xt\_zt[n*M + n_1]]$, and the state transfer probability of multi-component system is $xtl*m + n$.

6.3.4 Multi-component Maintenance Decision Model Without Considering Dependencies

In accordance with Behrman's optimal iterative equation [23], the decision model for state maintenance costs or the iterative equation for a train multi-component system is deduced:

$$v_N(l) + \rho_{\pi系}\left(T(l) + t_N\right) = C_N(l) + \sum_{n=1}^{5^M} xt[l*m+n] \times v^*_{N-1}(n) \tag{6.17}$$

In formula,

$v_N(l)$ indicates the state of train multi-component system as l, and the relative maintenance cost per unit time of the system when the equation is iterated to the Nth time.

$\rho_{\pi\,\text{system}}$ demonstrates the long-term unit time maintenance cost of train multi-component system.

$T(l)$ denotes the system maintenance cycle on the condition that the status of train multi-component system is l, which is also selected from the maintenance cycle set $T = \{180, 360, 540, 720, 1080\}$. Besides, the component maintenance cycle is consistent with that of train multi-component system.

t_N symbolizes the maintenance time of train multi-component system after equation iteration to the Nth time, and is the sum of maintenance time of all components following equation iteration to the Nth time.

$C_N(l)$ represents the state of train multi-component system as l, and the total maintenance cost at the stage of equation iteration to the Nth time.

$v_{N-1}{*}(n)$ indicates the train multi-component system state as R, and the minimum relative time unit cost of the system when the equation is iterated to step $N-1$.

The cost model for the pilot maintenance strategy of a train multi-component system does not incorporate dependency; instead, it straightforwardly aggregates the maintenance cost of each component. Consequently:

$$C_N(l) = \sum_{n_1=1}^{M} C_N^{(n_1)}(i, R(i), T(l)) \tag{6.18}$$

In formula,

$C_N^{(n1)}(i, R(i), T(l))$ indicates that when the state of train multi-component system is l, the state of part n_1 is i, and the maintenance mode $R(i)$ and maintenance cycle $T(l)$ are selected. Moreover, the equation is iterated to the total maintenance cost at the stage N.

The expression for $C_N^{(n_1)}(i, R(i), T(l))$ is as follows:

$$\begin{aligned} C_N^{(n_1)}(i, R(i), T(l)) &= C_M^{(n_1)}(i, R(i)) + C_T^{(n_1)} + C_S^{(n_1)} + C_{MV}^{(n_1)}(i, R(i), T(l)) \\ &= C_M^{(n_1)}(i, R(i)) + C_T^{(n_1)} + C_S^{(n_1)} \\ &+ \sum\nolimits_{j=1}^{5} P^{(n_1)}(R(i), i, j) \\ &\times \left[\sum\nolimits_{k=1}^{5} P^{(n_1)}(1, j, k) \times \left(T(l) + t_{R(i)}^{(n_1)}\right) \times \rho^{(n_1)}(i)\right] \\ &+ \sum\nolimits_{k=1}^{5} P^{(n_1)}(R(i), i, j) \times \left[P^{(n_1)}(1, j, 5) \times t_{R(i)}^{(m_1)} \times C_F\right] \end{aligned} \tag{6.19}$$

In formula,

$C_M^{(n_1)}(i, R(i))$ indicates the direct maintenance cost of part n_1;

$C_T^{(n_1)}$ symbolizes the testing cost of part $n_{1;}$

$C_S^{(n_1)}$ demonstrates the disassembly cost of part $n_{1;}$

$C_{MV}^{(n_1)}(i, R(i), T(l))$ indicates the risk cost of part n_1;

$\rho^{(n_1)}(i)$ shows the random failure maintenance cost of part n_1;

$t_{R(i)}^{(n_1)}$ illustrates the maintenance time required for part n_1 to adopt maintenance mode $R(i)$;

$P^{(n_1)}(1, j, k)$ denotes the probability of sudden deterioration failure after part n_1 is transferred to j state, $j < k$,

$P(n_1)(1, j, 5)$ indicates the probability of sudden deterioration failure after part 1 is transferred to j state, $j < 5$.

6.3.5 Component Dependency Modeling

Component dependency modeling is the foundation of state maintenance strategy optimization of train multi-component systems considering the dependency. Component dependency refers to maintenance dependency arising from correlation (interaction) between components, including economic dependency, structural dependency as well as fault dependency.

Due to the interdependent relationship between components, system maintenance costs can be reduced by using opportunistic group maintenance. Economic interdependence factors in the overlapping maintenance activities among components, along with the costs associated with disassembly and installation. Structural dependence takes into consideration the reduction in maintenance time and the mitigation of downtime losses for multi-component systems. Fault dependence accounts for the mutual influence of faults among components, aiming to diminish the state transition probability of multi-component systems.

The mutual influence diagram of faults between components can be represented by a fault chain [26]. Furthermore, the degree of fault correlation between components is termed the fault correlation coefficient. In this research, the fault correlation coefficient matrix between components is defined as follows:

$$\begin{matrix} 1 & 2 & \ldots M \end{matrix}$$
$$\begin{bmatrix} 1 & K_{12} & \ldots & K_{1M} \\ K_{21} & K_{22} & \ldots & K_{2M} \\ \ldots & \ldots & \ldots & \ldots \\ K_{M1} & K_{M2} & \ldots & 1 \end{bmatrix} \tag{6.20}$$

In formula,

$K_{n_1,n_1'}$ indicates the probability that a component n_1 failure leads to a component n_1' failure $\left(n_{1,}n_1' = 1, 2, ..., M\right)$.

$K_{n_1,n_1'}$ can be expressed as follows:

$$K_{\mathrm{ran,h}} = \frac{K_{r_1}|\mathrm{n}_1}{K_{rr_1}} \tag{6.21}$$

In formula,

$K_{n_1'}$ indicates the total failure frequency of component n_1';

$K_{n_1'}|n_1$ indicates the number of times that component n_1' fails when component n_1 fails. In instances where the available data is insufficient, the probability is assigned based on maintenance experience.

6.3.6 Multi-component Maintenance Decision Model Considering Dependency

The adjustment of the state maintenance cost decision model of the train multi-component system under the condition of one, two, and three dependencies is studied and the corresponding maintenance cost model is introduced as follows:

(1) There is a kind of dependence

(1) Economic dependence

In this research, the state maintenance cost decision model of a train multi-component system considering economic dependence is to adjust the disassembly cost in the trial maintenance strategy cost model of a multi-component system, and the state maintenance cost model of a train multi-component system considering economic dependence (presuming the existence of S economically dependent parts):

$$
\begin{aligned}
C_N(l) = {} & \sum\nolimits_{n_1=1}^{M} C_M^{(n_1)}(i, R(i)) + \sum\nolimits_{n_1=1}^{M} C_T^{(n_1)} + C_S^S + C_S^{M-S} \\
& + \sum\nolimits_{n_1=1}^{M} \sum\nolimits_{j=1}^{5} P^{(n_1)}(R(i), i, j) \\
& \times \left[\sum\nolimits_{k=1}^{5} P^{(n_1)}(1, j, k) \times \left(T(l) + t_{R(i)}^{(n_{p2})} \right) \times \rho^{(n_1)}(i) \right] \\
& + \sum\nolimits_{n_1=1}^{M} \sum\nolimits_{j=1}^{5} P^{(n_1)}(R(i), i, j) \times \left[P^{(n_1)}(1, j, 5) \times t_{R(i)}^{(m_2)} \times C_F \right]
\end{aligned}
\tag{6.22}
$$

In formula,

C_S^S indicates the disassembly cost of S economically dependent components, and is the maximum disassembly cost among S components;

C_S^{M-S} illustrates the disassembly and assembly cost of M-S components without economic dependence, and is the sum of the disassembly and assembly cost of M-S components.

(2) Structural dependence

In this research, the decision model for state maintenance costs in a train multi-component system, considering structural dependence, primarily takes into account the reduction of downtime costs within the trial maintenance strategy cost model for a multi-component system through group maintenance and the reduction of maintenance time within the decision model for state maintenance costs in a multi-component system. State maintenance cost model of train multi-component system considering structural dependence (assuming Q components with structural dependence):

$$
\begin{aligned}
C_N(l) = & \sum_{n_1=1}^{M} C_M^{(n_1)}(i, R(i)) + \sum_{n_1=1}^{M} C_T^{(n_1)} + \sum_{m_1=1}^{M} C_S^{(n_1)} \\
& + \sum_{n_1=1}^{M} \sum_{j=1}^{5} P^{(n_1)}(R(i), i, j) \\
& \times \left[\sum_{k=1}^{5} P^{(n_1)}(1, j, k) \times \left(T(l) + t_{R(i)}^{(n_2)} \right) \times \rho^{(n_1)}(i) \right] \\
& + \sum_{n_3=1}^{M-Q} \sum_{j=1}^{5} P^{(n_3)}(R(i), i, j) \times \left[P^{(n_3)}(1, j, 5) \times t_{R(i)}^{(n_3)} \times C_F \right] \\
& + \sum_{n_4=1}^{Q} \sum_{j=1}^{5} \left\{ \prod_{n_5=1}^{Q} P^{(n_5)}(R(i), i, j) \times \left[P^{(n_j)}(1, j, 5) \times t_{R(i)max}^{Q} \times C_F \right] \right\}
\end{aligned}
\tag{6.23}
$$

In formula,

$M-Q$ indicates the number of components that are not structurally dependent;

$t_{R(i)max}^{Q}$ illustrates that at the Nth iteration of the equation, Q components with structural dependence take the maximum maintenance time aligned with the maintenance mode $R(i)$.

Among them:

$n_1 = 1, 2, \ldots, M$, M indicates the number of train multi-component system components;

$n_3 = 1, 2, \ldots, M - Q$ demonstrates that Part n_3 is included in $M-Q$;

$n_4, n_5 = 1, 2, \ldots, Q$ symbolizes that Parts n_4 and n_5 are included in Q.

The multi-component system maintenance time t_N of the state maintenance cost decision model of train multi-component system is adapted to the cumulative maintenance time of $M-Q$ components without structural dependence plus the maximum maintenance time $t_{R(i)max}^{Q}$ of Q components with structural dependence.

(3) Fault dependence

In this research, the adjustment of the state transition probability of the multi-component system in the decision-making model of the state maintenance cost of the train multi-component system by group maintenance is predominantly considered. In the state maintenance cost model of the train multi-component system, fault dependency is taken into account. In this context, the paper considers fault dependency only between the faulty component and one other component, omitting consideration of weak fault dependency between the faulty component and other components.

$$
\begin{aligned}
C_N(l) = & \sum_{M}^{M} C_M^{(n_1)}(i, R(i)) + \sum_{n_1=1}^{M} C_T^{(n_1)} + \sum_{n_1=1}^{M} C_S^{(n_1)} \\
& + \sum_{n_1=1}^{M} \sum_{j=1}^{5} P^{(n_1)}(R(i), i, j) \\
& \times \left[\sum_{k=1}^{5} P^{(n_1)}(1, j, k) \times \left(T(l) + t_{R(i)}^{(n_{p2})} \right) \times \rho^{(n_1)}(i) \right]
\end{aligned}
$$

$$+\sum_{n_1=1}^{M}\sum_{j=1}^{5} P^{(n_1)}(R(i),i,j)\times\left[P^{(n_1)}(1,j,5)\times t_{R(i)}^{(n_j)}\times C_F\right] \tag{6.24}$$

In this research, fault chain is used to construct the relationship model of fault dependence, and fault correlation coefficient is employed to construct the state influence matrix [27]. It is presumed that components n_1 and n_1' in a train multi-component system have fault dependence. Moreover, the state influence matrix arising from the fault of component n_1' on the fault of component n_1 is denoted as K_1, subsequently the state influence matrix is a matrix with five rows and five columns:

$$K_1=\begin{bmatrix} 1-K_{n_1'n_1} & 0 & 0 & 0 & K_{n_1'n_1} \\ 0 & 1-K_{n_1'n_1} & 0 & 0 & K_{r_1'n_1} \\ 0 & 0 & 1-K_{n_1'n_1} & 0 & K_{n_1'n_1} \\ 0 & 0 & 0 & 1-K_{n_1'n_1} & K_{n_1'n_1} \\ 0 & 0 & 0 & 0 & 0 \end{bmatrix} \tag{6.25}$$

When component n_1' fails, $xt\_zt\left[l{*}M+n_1'\right]=5$, part of the multi-component system state transition probability array $xt[l{*}m+n]$, which may result in component n_1 failure, is modified as follows:

$$xt[l*m+n]=\begin{cases} K_{n_1'n_1}\times dy[n_1',R(i)][[5*5+xt\_zt[n*M+n_1']] \\ \times dy[n_1,R(i)][5*5+xt\_zt[n*M+n_1]] \\ \times\prod_{n_6}^{M-2} dy[n_6,R(i)][xt\_zt[l*M+n_6]*5+xt\_zt[n*M+n_6]], \\ \text{当}xt\_zt[l*M+n_1]\neq xt\_zt[n*M+n_1] \\ K_{n_1'n_1}\times dy[n_1',R(i)][[5*5+xt\_zt[n*M+n_1']] \\ \times dy[n_1,R(i)][5*5+xt\_zt[n*M+n_1]] \\ \times\prod_{n_6}^{M-2} dy[n_6,R(i)][xt\_zt[l*M+n_6]*5+xt\_zt[n*M+n_6]] \\ +(1-K_{n_1'n_1})\times dy[n_1',R(i)][[5*5+xt\_zt[n*M+n_1']] \\ \times\prod_{n_6}^{M-2} dy[n_6,R(i)][xt\_zt[l*M+n_6]*5+xt\_zt[n*M+n_6]], \\ \text{当}xt\_zt[l*M+n_1]=xt\_zt[n*M+n_1] \end{cases} \tag{6.26}$$

Among them:
$n_6=1,2,\ldots M-2$, $M-2$ indicates all parts except parts n_1 and n_1'.

(2) There are two kinds of interdependence

(1) Economic dependence and structural dependence

The state maintenance cost model of a train multi-component system, considering both economic dependence and structural dependence, is formulated under the assumption of S economically dependent components and Q structural dependent components.

$$
\begin{aligned}
C_N(l) = & \sum_{n_1=1}^{M} C_M^{(n_1)}(i, R(i)) + \sum_{n_1=1}^{M} C_T^{(n_1)} + C_S^S + C_S^{M-S} \\
& + \sum_{n_1=1}^{M} \sum_{j=1}^{5} P^{(n_1)}(R(i), i, j) \\
& \times \left[\sum_{k=1}^{5} P^{(n_3)}(1, j, k) \times \left(T(l) + t_{R(i)}^{(n_i)} \right) \times \rho^{(n_1)}(i) \right] \\
& + \sum_{n_3=1}^{M-Q} \sum_{j=1}^{5} P^{(n_3)}(R(i), i, j) \times \left[P^{(n_3)}(1, j, 5) \times t_{R(i)}^{(n_3)} \times C_F \right] \\
& + \sum_{n_4=1}^{Q} \sum_{j=1}^{5} \left\{ \prod_{n_5=1}^{Q} ?P^{(n_5)}(R(i), i, j) \times \left[P^{(n_4)}(1, j, 5) \times t_{R(i)max}^{Q} \times C_F \right] \right\}
\end{aligned}
\tag{6.27}
$$

Additionally, the maintenance time of the train multi-component system is adjusted in the state maintenance cost decision model of the train multi-component system. Furthermore, the adjustment method is demonstrated in the state maintenance cost decision model of a train multi-component system considering structural dependence.

(2) Economic dependence and fault dependence

The state maintenance cost model for a train multi-component system, incorporating economic dependence and fault dependence (with the assumption of S economically dependent components), is equivalent to the state maintenance cost model for a train multi-component system considering economic dependence (Eq. 6.20). In addition, some elements of the multi-component system state transition probability in the decision-making model of the train multi-component system state maintenance cost are modified, and the specific correction method is demonstrated in Eq. (6.24).

(3) Structural dependence and fault dependence

The state maintenance cost model of a train multi-component system considering structural dependence and fault dependence (presuming Q components with structural dependence) is the same as the state maintenance cost model of a train multi-component system considering structural dependence (6.21). Moreover, the multi-component system maintenance time of the multi-component system state maintenance cost decision model was adjusted and certain components of the probability governing the system's state transitions underwent alterations. The adjustment method was demonstrated in the structural dependence considering the train multi-component system state maintenance cost decision model, and the correction method was demonstrated in Eq. (6.24).

(3) There are three kinds of dependencies
The state maintenance cost model of a train multi-component system considering economic dependence, structural dependence, and fault dependence (assuming S economically dependent components and Q structurally dependent components) is the same as the state maintenance cost model of a train multi-component system considering economic dependence and structural dependence (6.25). Furthermore, adjustments were made to the maintenance duration within the decision model for the state maintenance cost of the multi-component system, and certain aspects of the probability governing the system's state transitions underwent modifications. The adjustment method was demonstrated in the structural dependence considering the train multi-component system state maintenance cost decision model, and the correction method was demonstrated in Eq. (6.24).

6.3.7 Objective Function of State Maintenance Strategy Optimization for a Multi-component System

Given the series relationship among the components in the train multi-component system, the reliability of the entire system is the product of the reliability values of all individual components within the system.

$$R_l(t) = \prod_{n_1=1}^{M} (1 - dy[n_1, R(i)][xt\_zt[l{*}M + n_1] * 5 + 5]) \tag{6.28}$$

In formula, $R_l(t)$ indicates the system reliability when the train multi-component system state is l.

In addition, $R_l(t) \geq R_{\text{system}}$, R_{system} represents the reliability limit to ensure the normal operation of the train multi-component system (0.8).

The availability of a train multi-component system is defined as the ratio of the uninterrupted running time of the multi-component system to the total actual running time [2]:

$$\boldsymbol{A}_{\text{system}} = 1 - \frac{\sum t_N}{\sum T(l) + \sum t_N} \tag{6.29}$$

In formula, $\sum T(l)$ indicates the sum of maintenance cycles within the running time of the train multi-component system;

$\sum t_N$ indicates the sum of maintenance time during the running time of the train multi-component system.

Additionally, $\boldsymbol{A}_{\text{system}} \geq \boldsymbol{A}_{\text{system0}}$, $\boldsymbol{A}_{\text{system0}}$ indicates the utilizable limit to ensure the normal operation of the train multi-component system, which is 0.9.

The system maintenance cost of the train multi-component system does not exceed its replacement cost during each maintenance event. The cost constraint for the replacement of the train multi-component system is articulated as follows:

$$\rho_{\pi\text{system}}(T(l) + t_N) \leq v_{g\text{system}} \tag{6.30}$$

In formula,

$v_{g\text{system}}$ indicates total cost of train multi-component system replacement.

By integrating the maintenance cost decision model, reliability model, and availability model of the multi-component train system, the optimization model for the state maintenance strategy of the train multi-component system is derived as follows:

$$\begin{cases} v_N(l) + \rho_{\pi\text{system}}(T(l) + t_N) = C_N(l) + \sum\limits_{n=1}^{5^M} xt[l*m + n] \times v_{N-1}{}^*(n) \\ R_l(t) \geq R_{\text{system}} \\ A_{\text{system}} \geq A_{\text{system0}} \\ \rho_{\pi\text{system}}(T(l) + t_N) \leq v_{g\text{system}} \end{cases} \tag{6.31}$$

6.4 Algorithm

In this research, the solution algorithm of the condition maintenance strategy optimization model adopts the strategy iteration method, and the solution process [17] is demonstrated in Fig. 6.4. There are two specific cases as follows.

6.4.1 Single Component Maintenance Decision Optimization Algorithm

Step 1: Select the initial policy:

The maintenance strategy π of A single train component contains $A(i)$ maintenance mode and maintenance period selected by each state, which can be expressed as

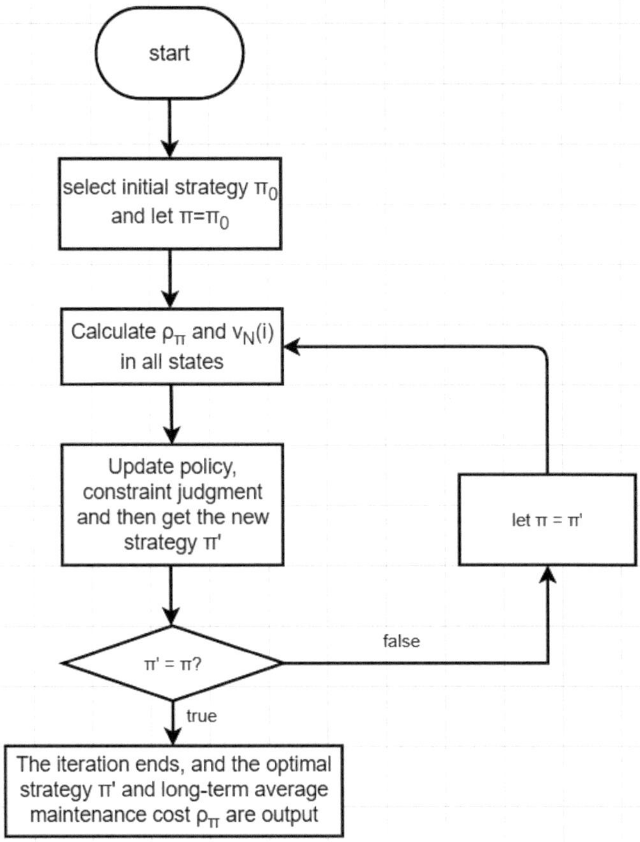

Fig. 6.4 Flowchart of the policy iteration algorithm

$\pi = \{A(1), A(2), \ldots\ldots., A(N-1), A(N)\}$, where $A(i)$ is the combination of maintenance mode and maintenance period under the single component state i, and A can be expressed as $A(i) = \{R(i), T(i)\}$. All the maintenance strategies π of single train component constitute the maintenance strategy set of single train component, and the initial strategy π_0 can be arbitrarily selected from the maintenance strategy set for a single train component.

Step 2: Calculate ρ_π and $v_N(i)$:

Allow the relative maintenance cost per unit time of a single train component under state 1 to be 0, that is:

$$v_N(1) = 0 \tag{6.32}$$

Solve the linear equations in five states, namely, formula (6.10), and get ρ_π and $v_N(i)$ under the selected strategy.

Step 3: Update the policy:

Various maintenance cycles and methodologies are employed for each state to compute the value of Formula (6.31).

$$C_N(i, R(i), T(i)) - \rho_\pi\left(T(i) + t_{R(i)}\right) + \sum_{j=1}^{5} P(R(i), i, j) \times v_{N-1}{}^*(j) \tag{6.33}$$

In comparison to the value under the strategy selected in Step 2, and when the reliability, availability, and replacement cost of a single train component satisfy the constraint conditions, the maintenance strategy π' that minimizes ρ_π can be obtained.

Step 4: Iteration terminates:

When $\pi' = \pi$, the maintenance strategy π' is the optimal maintenance strategy for single train component. Otherwise, set $\pi' = \pi$ and repeat steps 2–3 until π' is the same as π.

Step 5: Add the counting function to the above process:

N1 increments with each minor repair of a single train component. N2 is tallied once for each type of condition repair of a single train component. N3 is registered for each type 2 condition repair of a single train component. N4 is accrued with each major overhaul of a single train component.

6.4.2 Multi-component System Maintenance Decision Optimization Algorithm

In the optimization model of the condition maintenance strategy for a train multi-component system, the application process of the strategy iteration method exhibits a slight deviation from that employed in solving the optimization model for the condition maintenance strategy of a single train component.

(1) Distinct initial strategies

Assume that maintenance policy $\pi_{\text{system}l} = \{A_1(l), A_2(l), \ldots\ldots A_{M-1}(l), A_M(l)\}$ in multi-component system state l and maintenance policy of component M in multi-component system state l are adopted in $A_M(l)$. Besides, the maintenance strategy $\pi_{\text{system}l}$ of train multi-component system in all states constitutes the maintenance strategy set of train multi-component system, and the initial strategy $\pi_{0\text{system}}$ can be arbitrarily selected from the maintenance strategy set of train multi-component system.

(2) Distinct target values

The target value changes from the long-term unit time cost of the component ρ_π to the long-term unit time cost of the multi-component system $\rho_{\pi \text{system}}$.

(3) Value iteration equation is distinct

Value iteration equation is changed to:

$$C_N(l) - \rho_{\pi \text{system}}(T(l) + t_N) + \sum_{n=1}^{5^M} xt[l^*M + n] \times v_{N-1}{}^*(n) - v_N(l) \tag{6.34}$$

(4) Distinct constraints

The constraints are modified to encompass the reliability, availability, and replacement cost constraints of the train multi-component system.

6.5 Case Verification

The optimization model for the condition maintenance strategy of the train multi-component system focuses on the wheelset and axle box device within the running subsystem as the research object. The wheelset and axle box device are predominantly composed of three key components: axle box bearing, axle box body, and wheelset. The research object of the optimization model of condition maintenance strategy of single train components is axle box bearing. Example verification comprises three integral components:

(1) Determination of relevant parameters of axle box bearing, axle box body, and wheelset in the optimization model of maintenance strategy.
(2) Analysis of optimal maintenance strategy for axle box bearing.
(3) Examination of optimal maintenance strategy for wheelset axle box device.

6.5.1 Case Description

Based on the field investigation data, the relationships among the components of axle box bearing, axle box body, and wheelset, along with the optimization distribution parameters for maintenance, maintenance time, maintenance cost, service life, and the replacement cost of the wheelset axle box device, are established.

(1) Interdependence between axle box bearing, axle box body, and wheel set components
Economic dependence: The three components are economically dependent on each other.

Structural dependence is observed in the maintenance processes: the axle box body and axle box bearing require unloading during wheelset maintenance, the axle box bearing requires unloading during the physical examination and repair of the axle box, and the axle box body requires unloading during the maintenance of the axle box bearing. Specifically, structural dependence between the wheelset and the axle box body, as well as the axle box bearing, should be considered when the wheelset maintenance mode R = 3 or 4. Similarly, when the axle box undergoes physical examination repair mode R = 3 or 4, and the axle box bearing undergoes repair mode R = 3 or 4, the structural dependence between the axle box body and the axle box bearing should be taken into account.

Fault dependence refers to the causal propagation of failures between components. For instance, axle box bearing failure may lead to axle box body failure, which can occur under poor bearing operating conditions, axle shaft channeling, or impact wear on the end cover of the axle box body. In turn, axle box body failure may result in wheelset failure, including axle box body cracks that cause lateral instability of the wheelset, increase wheel–rail lateral forces, and accelerate wheelset wear. Utilize the fault chain concept to delineate the interdependency of faults among the axle box bearing, axle box body, and wheelset components, as illustrated in Fig. 6.5.

(2) Determination of relevant parameters of the model.

(1) Fault deterioration distribution parameters (Table 6.3).

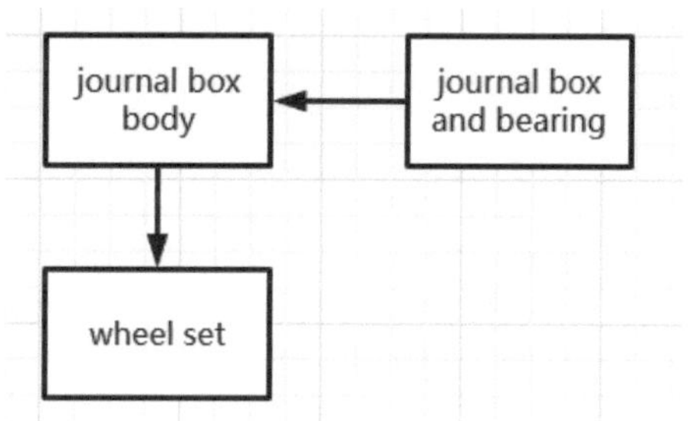

Fig. 6.5 Fault chain diagram of key components of wheelset axle box device

Table 6.3 Estimated fault distribution parameters of axle box bearings, axle box bodies, and wheelsets

Component	Estimated value of shape parameter β_1	Estimated value of scale parameter η_1
Axle box bearing	6.5	326
Wheel set	7.1	367
Axle box body	7.3	358

Table 6.4 Distribution parameters of axle box bearing maintenance optimization

Maintenance mode	Maintenance optimization distribution parameters β_2	Maintenance distribution optimization parameter η_2
R = 1	–	–
R = 2	1.3	526
R = 3	2.5	493
R = 4	–	–

Table 6.5 Distribution parameters of axle box body repair optimization

Maintenance mode	Maintenance optimization distribution parameters β_2	Maintenance distribution optimization parameter η_2
R = 1	–	–
R = 2	3.8	543
R = 3	1.7	452
R = 4	–	–

Table 6.6 Distribution parameters of wheel set maintenance optimization

Maintenance mode	Maintenance optimization distribution parameters β_2	Maintenance distribution optimization parameter η_2
R = 1	–	–
R = 2	3.9	560
R = 3	4.5	434
R = 4	–	–

(2) Distribution parameters of maintenance optimization

Assume that bearing maintenance optimization distribution parameters are demonstrated in Table 6.4, axle box maintenance optimization distribution parameters are illustrated in Table 6.5, and wheelset maintenance optimization distribution parameters are indicated in Table 6.6.

(3) Maintenance time (Table 6.7).

(4) Maintenance cost.

Assume that the direct maintenance cost, detection cost, daily maintenance cost, and disassembly cost of axle box bearing is demonstrated in Tables 6.8, 6.11, 6.14, and 6.17, respectively, whereas the direct maintenance cost, detection cost, daily maintenance cost, and disassembly cost of axle box are demonstrated in Tables 6.9, 6.12, 6.15, and 6.18. The direct maintenance cost, detection cost, daily maintenance cost, and disassembly cost of wheelsets are demonstrated in Tables 6.10, 6.13, 6.16, and 6.19.

Table 6.7 Lists the maintenance times for bearings, axle boxes, and wheelsets in distinct maintenance modes

Maintenance mode	Axle box bearing	Axle box body	Wheel set
R = 1	2	2	4
R = 2	16	8	16
R = 3	40	24	48
R = 4	56	48	56

Table 6.8 Cost of direct maintenance of axle box bearings

Project	R = 1	R = 2	R = 3	R = 4
S = 1	120	300	700	8000
S = 2	120	350	800	8000
S = 3	120	400	900	8000
S = 4	120	450	1000	8000
S = 5	120	500	1100	8000

Table 6.9 Cost of direct maintenance of axle box bodies

Project	R = 1	R = 2	R = 3	R = 4
S = 1	150	3000	4000	11,080
S = 2	150	3000	4000	11,080
S = 3	150	4000	5000	11,080
S = 4	150	4000	4500	11,080
S = 5	150	5000	5000	11,080

Table 6.10 Cost of direct maintenance of wheel sets

Project	R = 1	R = 2	R = 3	R = 4
S = 1	110	2000	4000	15,367
S = 2	110	3000	5000	15,367
S = 3	110	3500	6000	15,367
S = 4	110	4000	7000	15,367
S = 5	110	5000	8000	15,367

Table 6.11 Testing costs of axle box bearings

Project	R = 1	R = 2	R = 3	R = 4
Testing cost	0	100	500	1000

Table 6.12 Testing costs of axle box bodies

Project	R = 1	R = 2	R = 3	R = 4
Testing cost	50	200	400	700

Table 6.13 Testing costs of wheel sets

Project	R = 1	R = 2	R = 3	R = 4
Testing cost	50	50	500	500

Table 6.14 Daily maintenance costs of axle box bearings

Project	S = 1	S = 2	S = 3	S = 4	S = 5
Daily maintenance cost	50	50	100	150	200

Table 6.15 Daily maintenance costs of axle box bodies

Project	S = 1	S = 2	S = 3	S = 4	S = 5
Daily maintenance cost	50	50	100	150	200

Table 6.16 Daily maintenance costs of wheel sets

Project	S = 1	S = 2	S = 3	S = 4	S = 5
Daily maintenance cost	50	50	100	150	200

(3) The axle box bearing, axle box body, and wheelset are numbered 1, 2, and 3, respectively. According to the above mathematical statistics, the probability K12 of the axle box bearing leading to the failure of the axle box body is 0.3, and the probability K23 of the wheelset failure arising from the axle box body is 0.35.
(4) The life of the axle box bearing is set to 10 years, the life of the axle box body is set to 6 years, and the life of the wheelset is set to 7.5 years.
(5) The replacement cost of the axle box bearing is 9000 yuan, and the replacement cost of the wheelset axle box device is set at 40,000 yuan.

Table 6.17 Cost of removing and assembling axle box bearings

Project	R = 1	R = 2	R = 3	R = 4
Cost of removing and assembling	0	0	300	500

Table 6.18 Cost of removing and assembling axle box bodies

Project	R = 1	R = 2	R = 3	R = 4
Cost of removing and assembling	0	0	500	1000

Table 6.19 Cost of removing and assembling wheel sets

Project	R = 1	R = 2	R = 3	R = 4
Cost of removing and assembling	0	0	500	1500

6.5.2 Analysis of Optimization Results of Axle Box Bearing Maintenance Strategy

The maintenance data related to axle box bearings were substituted into formula (6.12) and programmed with Python. The iteration time was 34 s, and the optimal maintenance strategy of axle box bearings was obtained, as demonstrated in Table 6.20. The long-term average maintenance cost of axle box bearings was 1.34 yuan/hour.

As evident from Table 6.20, when the axle box bearing is in status 1, it is in optimal condition and operates under normal conditions. The maximum maintenance period is 1080 days and the maintenance mode is minor repair. When the state of the axle box bearing is 2, the state is outstanding and it is a mild failure. Besides, the maintenance cycle is 720 days and the maintenance mode is minor repair, which is consistent with the nature of high reliability of the bearing. In the scenario where the axle box bearing is in state 3, a significant deterioration in its condition starts. The maintenance period is set at 720 days, and the maintenance mode involves a form of state repair. When the status of the axle box bearing is 4, the performance of the axle box bearing is extremely poor at this time, and the maintenance effort needs to be increased. The maintenance cycle is 540 days, and the maintenance mode is the second type of state repair. When the axle box bearing is in state 5, it has reached the "complete failure" condition. The maintenance plan for this scenario involves a maintenance period of 360 days, and the maintenance mode is an overhaul.

The maintenance period and maintenance mode in the optimization model of maintenance strategy for a single train component vary with the status of a single component. Moreover, the biggest difference from the traditional maintenance plan with a fixed period is that the maintenance period is variable. Following the maintenance period is fixed, the same axle box bearing data in the optimization model example of the maintenance strategy of a single train component is employed for calculation, and the comparison between the long-term average maintenance cost of axle box bearing under the fixed period is obtained, as demonstrated in Fig. 6.6.

As demonstrated in Fig. 6.6, compared with the long-term average maintenance cost of axle box bearings under a fixed maintenance period, the long-term average

Table 6.20 Optimal maintenance strategies for axle box bearings

State i	1	2	3	4	5
Maintenance mode R	1	1	2	3	4
Maintenance cycle T	1080	720	720	540	360

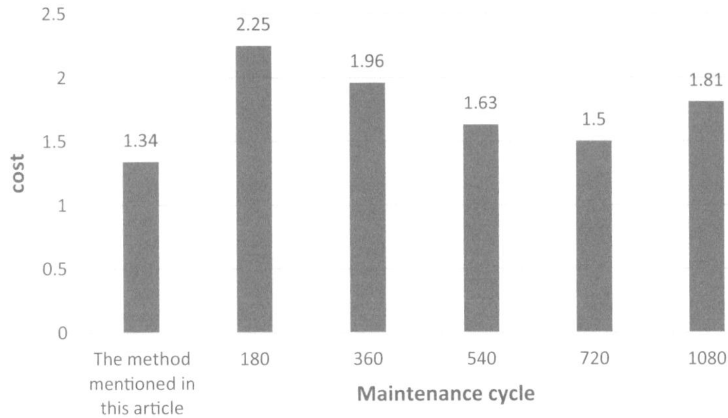

Fig. 6.6 Comparison of the long-term average maintenance cost of axle box bearings with that of fixed maintenance cycles

maintenance cost of axle box bearings under the optimal maintenance strategy in this paper is lower than the long-term average maintenance cost of axle box bearings under five kinds of fixed maintenance period. Due to the fact that the optimal maintenance strategy of axle box bearings in this paper will not lead to "over maintenance" when the condition of axle box bearings is outstanding choose the maintenance mode with larger maintenance period and less maintenance strength. If the axle box bearing is in a suboptimal condition, employing a maintenance strategy with a shorter maintenance period and higher maintenance intensity will not result in "insufficient maintenance." In this paper, the maintenance strategy of axle box bearing is optimized according to the above principles, and the long-term average maintenance cost of axle box bearing under the optimal maintenance strategy is the lowest.

When the maintenance cycle of the axle box bearing is fixed to 180 days or 360 days, the maintenance times are longer, and the reliability of the axle box bearing is high, resulting in the "overmaintenance" of the axle box bearing. The maintenance cycle of 180 days or 360 days is appropriate for the axle box bearing experiencing significant performance degradation: when the maintenance cycle of the axle box bearing is fixed at 1080 days, the maintenance opportunity of the axle box bearing may be missed, which is easy to result in to the "insufficient maintenance" of the axle box bearing. Besides, the maintenance cycle of 1080 days is suitable for the axle box bearing with very high reliability. On the condition that the maintenance cycle of axle box bearing is fixed at 540 days or 720 days, in comparison with the condition maintenance cost model of single train component in this paper, the risk cost of axle box bearing under the two maintenance cycles rises. The maintenance cycle of 540 days or 720 days is suitable for axle box bearings with high reliability. Consequently, for axle box bearings, the condition maintenance strategy, in which the maintenance period is adjusted based on the condition of axle box bearings, proves to be more economical and reliable.

6.5.3 Analysis of Optimization Results of Maintenance Strategy for Wheelset Axle Box Device

Maintenance data related to the axle box bearing, axle box body, and wheelset were utilized to validate the optimization model for the condition maintenance strategy of the train multi-component system considering dependency. Python was used for programming, and there were 55 iterations. Each iteration involved approximately 780,000 calculations, and the computation time for each iteration was about 18 min. Table 6.21 shows the optimal maintenance strategy (Fig. 6.7).

The maintenance period in the model of optimal maintenance strategy for wheelset axle box device considering the dependence is fixed, and the parameters in the example of optimization model of maintenance strategy for multi-component train system are used for calculation. Besides, the long-term operating unit time cost of the wheelset axle box device under a fixed period is obtained, as demonstrated in Fig. 6.8.

As demonstrated in Fig. 6.8, compared with the long-term average maintenance cost of wheelset axle box device under the fixed maintenance cycle, the average maintenance cost over the long term under the trial maintenance strategy is lower than that under the five fixed maintenance cycles. This is due to the fact the trial maintenance strategy has been optimized: on the condition that the parts of the finger wheelset axle box device are in a satisfactory condition, the maintenance mode with a larger maintenance period and smaller maintenance intensity is selected, which will not lead to "overmaintenance." When the components are in a deteriorated state, the maintenance mode with a shorter maintenance period and more extensive maintenance effort is chosen, avoiding "insufficient maintenance" and decreasing the long-term average maintenance cost of the wheelset axle box device.

In addition, it is evident from Fig. 6.8 that the long-term average maintenance cost under the condition of the maintenance strategy for the wheelset axle box device, considering dependency, is 4.54 yuan/hour. 0.54 yuan/hour lower than the long-term average maintenance cost under the trial maintenance strategy of wheelset axle box device. Besides, the condition maintenance strategy of the wheelset axle box device considering the dependence is optimized twice: the initial optimization is identical to the optimization of the trial maintenance strategy. The second optimization takes into account the interdependence of components, reducing the disassembly cost, maintenance time, shutdown loss cost and maintenance times of the wheelset axle box device, and accordingly reducing the long-term average maintenance cost of the wheelset axle box device. Consequently, the long-term average maintenance cost of wheelset axle box device under the condition maintenance strategy considering the dependency is lower.

In conclusion, the optimization model of the condition maintenance strategy of train multi-component system introduced by dependency is more cost-effective and dependable, and has certain reference significance for the formulation of maintenance strategy of rail transit train system.

Table 6.21 Optimal maintenance strategy for wheelset axle box device considering dependency

Multi-component system status	Axle box bearing (state, maintenance mode)	Axle box body (state, maintenance mode)	Wheel set (state, maintenance mode)	Maintenance cycle
1(1,1,1)	(1,1)	(1,1)	(1,1)	720
2(1,1,2)	(1,1)	(1,1)	(2,2)	720
3(1,1,3)	(1,1)	(1,1)	(3,2)	540
4(1,1,4)	(1,1)	(1,1)	(4,3)	360
5(1,1,5)	(1,1)	(1,1)	(5,4)	180
6(1,2,1)	(1,1)	(2,2)	(1,1)	540
7(1,3,1)	(1,1)	(3,3)	(1,1)	360
8(1,4,1)	(1,1)	(4,3)	(1,1)	360
9(1,5,1)	(1,1)	(5,4)	(1,2)	180
10(1,2,2)	(1,1)	(2,2)	(2,2)	540
11(1,2,3)	(1,1)	(2,2)	(3,2)	540
12(1,2,4)	(1,1)	(2,2)	(4,3)	360
13(1,2,5)	(1,1)	(2,2)	(5,4)	360
14(1,3,2)	(1,1)	(3,3)	(2,2)	360
15(1,4,2)	(1,1)	(4,3)	(2,2)	360
16(1,5,2)	(1,1)	(5,4)	(2,3)	180
17(1,3,3)	(1,1)	(3,3)	(3,2)	360
18(1,3,4)	(1,1)	(3,3)	(4,3)	360
19(1,3,5)	(1,1)	(3,3)	(5,3)	360
20(1,4,3)	(1,1)	(4,3)	(3,2)	360
21(1,5,3)	(1,1)	(5,4)	(3,3)	180
22(1,4,4)	(1,1)	(4,3)	(4,3)	360
23(1,4,5)	(1,1)	(4,3)	(5,3)	360
24(1,5,4)	(1,1)	(5,3)	(4,3)	180
25(1,5,5)	(1,1)	(5,4)	(5,4)	180
26(2,1,1)	(2,1)	(1,1)	(1,1)	720
27(2,1,2)	(2,1)	(1,1)	(2,2)	720
28(2,1,3)	(2,1)	(1,1)	(3,2)	540
29(2,1,4)	(2,1)	(1,1)	(4,3)	360
30(2,1,5)	(2,1)	(1,1)	(5,3)	180
31(2,2,1)	(2,1)	(2,2)	(1,1)	540
32(2,3,1)	(2,1)	(3,3)	(1,1)	360
33(2,4,1)	(2,1)	(4,3)	(1,1)	360
34(2,5,1)	(2,1)	(5,4)	(1,2)	180
35(2,2,2)	(2,1)	(2,2)	(2,2)	540

(continued)

Table 6.21 (continued)

Multi-component system status	Axle box bearing (state, maintenance mode)	Axle box body (state, maintenance mode)	Wheel set (state, maintenance mode)	Maintenance cycle
36(2,2,3)	(2,1)	(2,2)	(3,2)	540
37(2,2,4)	(2,1)	(2,2)	(4,3)	360
38(2,2,5)	(2,1)	(2,2)	(5,4)	360
39(2,3,2)	(2,1)	(3,3)	(2,2)	360
40(2,4,2)	(2,1)	(4,3)	(2,2)	360
41(2,5,2)	(2,1)	(5,4)	(2,3)	180
42(2,3,3)	(2,1)	(3,3)	(3,2)	360
43(2,3,4)	(2,1)	(3,3)	(4,3)	360
44(2,3,5)	(2,1)	(3,3)	(5,3)	360
45(2,4,3)	(2,1)	(4,3)	(3,2)	360
46(2,5,3)	(2,1)	(5,4)	(3,3)	180
47(2,4,4)	(2,1)	(4,3)	(4,3)	360
48(2,4,5)	(2,1)	(4,3)	(5,3)	360
49(2,5,4)	(2,1)	(5,3)	(4,3)	180
50(2,5,5)	(2,1)	(5,4)	(5,4)	180
51(3,1,1)	(3,2)	(1,1)	(1,1)	720
52(3,1,2)	(3,2)	(1,1)	(2,2)	720
53(3,1,3)	(3,2)	(1,1)	(3,2)	540
54(3,1,4)	(3,2)	(1,1)	(4,3)	360
55(3,1,5)	(3,2)	(1,1)	(5,4)	180
56(3,2,1)	(3,2)	(2,2)	(1,1)	540
57(3,3,1)	(3,2)	(3,3)	(1,1)	360
58(3,4,1)	(3,2)	(4,3)	(1,1)	360
59(3,5,1)	(3,2)	(5,4)	(1,2)	180
60(3,2,2)	(3,2)	(2,2)	(2,2)	540
61(3,2,3)	(3,2)	(2,2)	(3,2)	540
62(3,2,4)	(3,2)	(2,2)	(4,3)	360
63(3,2,5)	(3,2)	(2,2)	(5,4)	360
64(3,3,2)	(3,2)	(3,3)	(2,2)	360
65(3,4,2)	(3,2)	(4,3)	(2,2)	360
66(3,5,2)	(3,2)	(5,4)	(2,3)	180
67(3,3,3)	(3,2)	(3,3)	(3,2)	360
68(3,3,4)	(3,2)	(3,3)	(4,3)	360
69(3,3,5)	(3,2)	(3,3)	(5,3)	360
70(3,4,3)	(3,2)	(4,3)	(3,2)	360

(continued)

Table 6.21 (continued)

Multi-component system status	Axle box bearing (state, maintenance mode)	Axle box body (state, maintenance mode)	Wheel set (state, maintenance mode)	Maintenance cycle
71(3,5,3)	(3,2)	(5,4)	(3,3)	180
72(3,4,4)	(3,2)	(4,3)	(4,3)	360
73(3,4,5)	(3,2)	(4,3)	(5,3)	360
74(3,5,4)	(3,2)	(5,3)	(4,3)	180
75(3,5,5)	(3,2)	(5,4)	(5,4)	180
76(4,1,1)	(4,3)	(1,1)	(1,1)	540
77(4,1,2)	(4,3)	(1,1)	(2,2)	540
78(4,1,3)	(4,3)	(1,1)	(3,2)	540
79(4,1,4)	(4,3)	(1,1)	(4,3)	360
80(4,1,5)	(4,3)	(1,1)	(5,4)	360
81(4,2,1)	(4,3)	(2,2)	(1,1)	540
82(4,3,1)	(4,3)	(3,3)	(1,1)	360
83(4,4,1)	(4,3)	(4,3)	(1,1)	360
84(4,5,1)	(4,3)	(5,4)	(1,2)	180
85(4,2,2)	(4,3)	(2,2)	(2,2)	540
86(4,2,3)	(4,3)	(2,2)	(3,2)	540
87(4,2,4)	(4,3)	(2,2)	(4,3)	360
88(4,2,5)	(4,3)	(2,2)	(5,4)	360
89(4,3,2)	(4,3)	(3,3)	(2,2)	360
90(4,4,2)	(4,3)	(4,3)	(2,2)	360
91(4,5,2)	(4,3)	(5,4)	(2,3)	180
92(4,3,3)	(4,3)	(3,3)	(3,2)	360
93(4,3,4)	(4,3)	(3,3)	(4,3)	360
94(4,3,5)	(4,3)	(3,3)	(5,3)	180
95(4,4,3)	(4,3)	(4,3)	(3,2)	360
96(4,5,3)	(4,3)	(5,3)	(3,3)	180
97(4,4,4)	(4,3)	(4,3)	(4,3)	360
98(4,4,5)	(4,3)	(4,3)	(5,3)	180
99(4,5,4)	(4,3)	(5,3)	(4,3)	360
100(4,5,5)	(4,3)	(5,3)	(5,4)	180
101(5,1,1)	(5,4)	(1,2)	(1,1)	540
102(5,1,2)	(5,3)	(1,2)	(2,2)	360
103(5,1,3)	(5,3)	(1,2)	(3,2)	360
104(5,1,4)	(5,3)	(1,2)	(4,3)	360
105(5,1,5)	(5,3)	(1,2)	(5,3)	180

(continued)

Table 6.21 (continued)

Multi-component system status	Axle box bearing (state, maintenance mode)	Axle box body (state, maintenance mode)	Wheel set (state, maintenance mode)	Maintenance cycle
106(5,2,1)	(5,3)	(2,3)	(1,1)	360
107(5,3,1)	(5,3)	(3,3)	(1,1)	360
108(5,4,1)	(5,3)	(4,3)	(1,1)	360
109(5,5,1)	(5,3)	(5,4)	(1,2)	180
110(5,2,2)	(5,3)	(2,3)	(2,2)	360
111(5,2,3)	(5,3)	(2,3)	(3,2)	360
112(5,2,4)	(5,3)	(2,3)	(4,3)	360
113(5,2,5)	(5,3)	(2,3)	(5,3)	180
114(5,3,2)	(5,3)	(3,3)	(2,2)	360
115(5,4,2)	(5,3)	(4,3)	(2,2)	360
116(5,5,2)	(5,3)	(5,4)	(2,3)	180
117(5,3,3)	(5,3)	(3,3)	(3,2)	360
118(5,3,4)	(5,3)	(3,3)	(4,3)	360
119(5,3,5)	(5,3)	(3,3)	(5,3)	180
120(5,4,3)	(5,3)	(4,3)	(3,2)	360
121(5,5,3)	(5,3)	(5,4)	(3,3)	180
122(5,4,4)	(5,3)	(4,3)	(4,3)	360
123(5,4,5)	(5,3)	(4,3)	(5,3)	180
124(5,5,4)	(5,3)	(5,4)	(4,3)	180
125(5,5,5)	(5,3)	(5,4)	(5,4)	180

6.6 Summary

Commencing with the formulation of efficient maintenance strategies, this research establishes optimization models for the maintenance strategies of individual components and multiple-component systems. Besides, the models employ Markov processes to describe the state transition process of individual components and multiple-component systems, set the goal of minimizing long-term average maintenance costs, as well as optimize the maintenance methods and maintenance cycles with reliability, availability, and replacement costs as constraints.

In formulating the optimization model for the maintenance strategy of the multiple-component system, dynamic arrays are employed to store the system state transition probabilities, and the interdependence between components is taken into account. Finally, the maintenance strategy optimization process is transformed into a Markov decision problem, with the wheelset bearing device as the research object, and an optimal maintenance strategy solution method is presented.

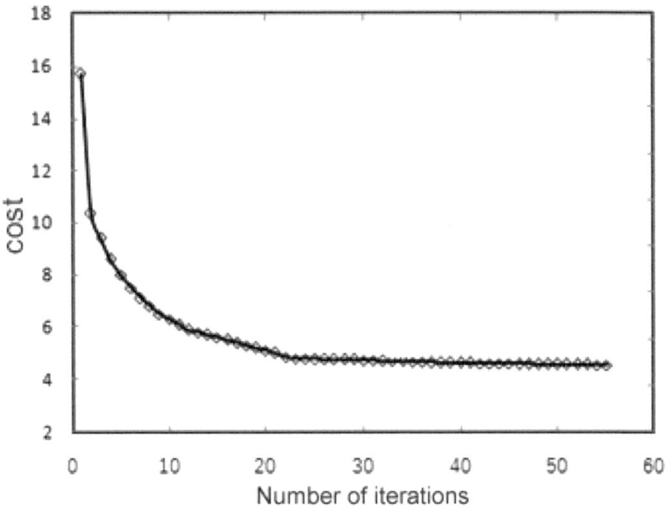

Fig. 6.7 Trend chart of long-term average maintenance cost of wheelset axle box device with the number of iterations

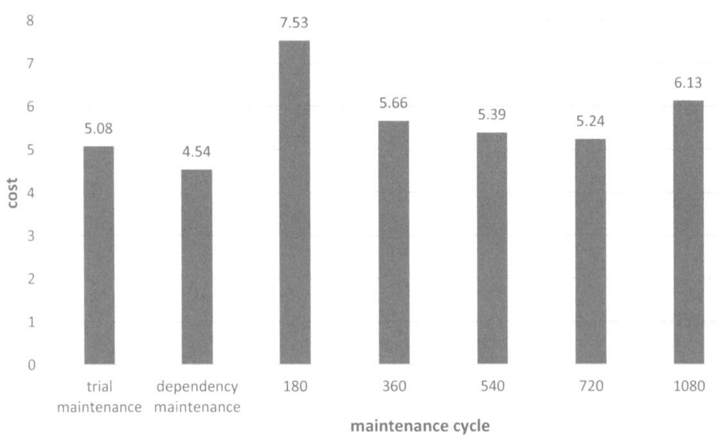

Fig. 6.8 Comparison of the long-term average maintenance cost of wheelset axle box under various maintenance strategies

References

1. General Administration of Quality Supervision, Inspection and Quarantine of the People's Republic of China. GB/T.21562-2008. Specifications and Examples for Reliability, Availability, Maintainability and Safety of Rail Transit. Beijing: Standardization Administration of China, 2008.(in Chinese)

2. Lijie, L. (2017). *Calculation of Reliability and Optimization of Operation and Maintenance Strategy for Rail Transit Train System Considering Mutual Influence among System Components*. Beijing: Beijing Jiaotong University. (in Chinese).
3. Christer, A. H., & Waller, W. M. (1984). Reducing production downtime using delay-time analysis. *The Journal of the Operational Research Society, 35*(6), 499–512.
4. Cox, D. (1972). Regression models and life tables. *Journal of the Royal Statistical Society: Series B (Methodological), 34.*
5. Song, G. (2017). *Preventive Maintenance Strategy for Several Types of Reliable Systems under Impact Model*. Southeast University. (in Chinese).
6. Jing-Yu, S., Wei, Z., & Bo, G. (2012). Decision of preventive maintenance and replacement of single-unit systems with gamma process. *Electronic Product Reliability & Environmental Testing.*
7. Wijnmalen, D. J. D., & Hontelez, J. A. M. (1992). Review of a Markov decision algorithm for optimal inspections and revisions in a maintenance system with partial information. *European Journal of Operational Research, 62*(1), 96–104.
8. Jingxuan, S. (2014). *Fault Prediction and Maintenance of Bogies for Urban Rail Transit Trains based on Reliability Analysis*. Beijing Jiaotong University. (in Chinese).
9. Ao, Z., Sheng, L., Ding, F., et al. (2017). Maintenance decision model for traction power supply equipment based on Markov process and its application. *Journal of the China Railway Society, 11*, 43–50. (in Chinese).
10. Wenjie, Z. (2019). *Research on Wheelset Wear Fault Prediction and Turning Maintenance Optimization for Metro Vehicles*. Shanghai Jiao Tong University. (in Chinese).
11. Thomas, L. (1986). A survey of maintenance and replacement models for maintainability and reliability of multi-item systems. *Reliability Engineering, 16*(4), 297–309.
12. Wang, H., Du, W., Liu, Z., et al. (2016). Maintenance of a multi-component system for high-speed trains considering joint failure and economic relevance. *Journal of Shanghai Jiaotong University, 50*(5), 660–667. (in Chinese).
13. Liu, Y. Study on degradation law of component status and failure correlation based on monitoring data. Southwest Jiaotong University. (in Chinese).
14. Pham, H., & Wang, H. (1996). Imperfect maintenance. *European Journal of Operational Research, 94*, 425–438.
15. He, D., Xiao, H., Yao, X., et al. (2019). Multi-objective optimization model and application of preventive maintenance for metro trains. *Journal of Guangxi University (Natural Science Edition), 44*(2), 9–15. (in Chinese).
16. Kali, X., Quan, C. (1999). Characteristic quantity of safety grade and its calculation method. *China Safety Science Journal.*
17. George, E., & Monahan, G. (1982). State of the art—a survey of partially observable Markov decision processes: Theory, models, and algorithms. *Management Science.*
18. Myers-Beaghton, A. K., & Vvedensky, D. D. (1999). Chapman-Kolmogrov equation for Markov models of epitaxial growth. *Journal of Physics A General Physics, 22*(11), L467.
19. Qian, J., Li, X., & Qin, W., et al. (2014). Prediction method of coal stacking time based on the mixed Gaussian hidden Markov model for belt conveyors. *Industry and Mine Automation, 40*(11), 26–30. (in Chinese).
20. Jiang, P., & Fan, G. (2015). Study on probability distribution properties of order statistics for three-parameter Weibull distribution. *Statistics and Decision.* (in Chinese).
21. Mou, M. (2017). Research on maintenance cost analysis method of subway vehicle life cycle. *China High-tech Enterprises.* (in Chinese).
22. Sheu, S. H., Chang, C. C., Chen, Y. L., et al. (2015). Optimal preventive maintenance and repair policies for multi-state systems. *Reliability Engineering & System Safety, 140*, 78–87.
23. Ye, P. (2012). Study on the optimization model of state maintenance strategy under incomplete maintenance. Tsinghua University. (in Chinese).
24. Xu, S., Geng, J., Wei, S., et al. (2018). Optimization of dynamic opportunistic group maintenance for series systems considering correlation. *Systems Engineering and Electronics.* (in Chinese).

25. Jin, X., Hong, Y., Zhang, M., et al. (2005). Fast calculation method for transition probability matrix of Markov model. *Journal of Projectiles, Rockets, Missiles and Guidance.* (in Chinese).
26. Wang, X., Shen, G., Zhang, Y., et al. (2015). Modeling the fault correlation coefficient of complex systems based on the fault chain. *Journal of Jilin University (Engineering and Technology Edition).* (in Chinese).
27. Sun, Y., Ma, L., & Mathew, J. (2009). Failure analysis of engineering systems with preventive maintenance and failure interactions. *Computers & Industrial Engineering, 57*(2), 539–549.

MIX
Papier aus verantwortungsvollen Quellen
Paper from responsible sources
FSC® C105338
FSC
www.fsc.org

If you have any concerns about our products,
you can contact us on
ProductSafety@springernature.com

In case Publisher is established outside the EU,
the EU authorized representative is:
Springer Nature Customer Service Center GmbH
Europaplatz 3, 69115 Heidelberg, Germany

Printed by Libri Plureos GmbH
in Hamburg, Germany